Why Construction Claims Occur and How to Prevent Them

Richard J. Long, P.E., P.Eng.

LONG INTERNATIONAL

Published in 2021 by Virtualbookworm.com Publishing Inc., College Station, Texas 77842, United States of America.

ISBN 978-1-63868-016-1

Library of Congress Number on file with the Publisher.

1 2 3 4 5 6 7 8 9 10

Table of Contents

Table of Contents

(continued)

Table of Contents
(continued)

Table of Contents
(continued)

Table of Contents
(continued)

Table of Contents
(continued)

Table of Contents
(continued)

Table of Contents
(continued)

Table of Contents
(continued)

Table of Contents

(continued)

Table of Contents
(continued)

Table of Contents

(continued)

Table of Contents
(continued)

Table of Contents

(continued)

List of Tables

Table of Contents
(continued)

List of Figures

Table of Contents
(continued)

Preface

In a perfect world, the owner never changes his mind, the engineer never alters his drawings, the contractor never malperforms, the resident engineer's decisions are perfect, and Mother Nature behaves herself. That perfect world does not exist.

In the real world, with geometric precision, the forces of owner, engineer, contractor, and Mother Nature combine to make change. The owner changes his mind. The engineer changes his drawings. The contractor fails to manage his job. Mother Nature then changes what the others have missed.

The high frequency of contract disputes must be recognized, and all reasonable measures undertaken by each party involved in today's engineering and construction projects. Technical complexities, the increasingly contentious nature of the contract environment, tight competitive bids, and owners with tough-minded and sometimes inequitable contractual attitudes have caused the final act of many construction projects to occur in an arbitration hearing or in the courtroom. Legal and expert costs for preparing and defending claims have exacerbated an already high-risk business environment for all construction contracting parties.

Although a construction project should be a mutually beneficial endeavor, adversarial relationships that can culminate in disputes are common. Owners and contractors are acutely aware of the significant monetary, labor, and equipment resources that they must commit to a construction project and the adverse effect any increase in these resource commitments would have on their bottom lines. Contractors and owners must be aware of and educated about claims and dispute resolution to protect their significant resource investment.

The possibility of being blindsided by a large claim has caused owners to question basic construction contract terms and conditions and the way they contract for design and build. Owners create exculpatory contract clauses to protect against the reoccurrence of such claims. When their workload is down, contractors take risks by accepting tough contract terms and conditions.

Owners expect more from their engineers, whether in-house or outside design professionals, but insist on lower costs and time to produce "issued for construction" documents. The engineering profession is continuously examining the quality of its product to determine if errors and omissions are increasing because of regression in professionalism or skill, or because of inadequate budgets and time for design work. If errors and omissions are increasing, how can they be better controlled? If they are not increasing, how can they be further reduced?

Professionals who prepare contracts and specifications are subjecting their work to continuous revisions to see if simplifications, exculpatory language, or new concepts are needed. Many defensive contract clauses such as "no damage for delay" and the usual subsurface exculpatory requirements to examine the site and not rely on borings have often failed as a result of improper contract administration or inadequate site investigations by the owner. Specifications are too often a "cut and paste" exercise that does not meet the varying needs of the project. Numerous claims could be avoided by knowledgeable and careful specification writing incorporating knowledge about construction law and claims.

Large value claims, when asserted successfully, may destroy the viability of a project or make construction more difficult by adding to the disenchantment between the owner and contractor. If a sizable claim is found to be valid, an owner faces serious problems. New financing may be required, later occupancy may be necessary, income from operation of the facility or rent from commercial property may be delayed or lost, and staffs may have to be funded longer than planned to administer the project.

The first five chapters of this book focus on the general topic, "Why Construction Claims Occur." Chapter 1 discusses the characteristics of a highly impacted project, the effect of claims on project costs, and the court's view of contract interpretation as it relates to the bases for contractor recovery for claims. Chapter 2 describes situations that cause costly claims. Chapter 3 describes why construction disputes occur and how to promote early an amicable settlement. Chapter 4 presents an assessment of problems associated with poor project management performance. Chapter 5 identifies typical problems leading to delays, cost overruns, and claims on process plant and offshore oil and gas projects.

The next eleven chapters of this book focus on the second general topic, "How to Prevent Construction Claims." Chapter 6 identifies claims mitigation steps that can be performed early in the project life cycle. Chapter 7 sets forth the concept of "Defining Project Requirements" to avoid project failure. Chapter 8 provides a detailed project checklist of questions that a senior-level project review team should use to ensure that all issues regarding the status of the project are covered during project reviews. Chapter 9 covers the essentials of performing a site investigation and provides detailed lists of information that need to be collected by the contractor to prepare its bid estimate. Chapter 10 covers the claims mitigation steps that should occur during construction by the owner's and contractor's project management teams. Chapter 11 describes the essential elements of cost control that need to be understood by the project manager. Change order management is discussed in Chapter 12. The keys to successful major project execution are set forth in Chapter 13 including the identification of life cycle stages of the project. Chapter 14 discusses project team motivators and demotivators that factor into determining the success of a project. Chapter 15 identifies how claims can be mitigated if the owner coordinates and cooperates on multi-prime construction projects. Commercial awareness training for EPC projects is discussed in Chapter 16.

The last three chapters present a summary of the previous discussions. A comprehensive list of risks that should be evaluated by the contractor as part of its due diligence before deciding to pursue a potential project is provided in Chapter 17. Chapter 18 presents a list of recommendations for owners and contractors for avoiding disputes. And finally, Chapter 19 presents a comprehensive checklist of the types of construction claims that arise during EPC projects, the documentation that is needed to support these types of claims, and the analyses that are typically performed to support claims related to schedule delay, quantities, man-hours, and costs.

Richard J. Long, P.E., P.Eng.

About the Author

Richard J. Long, P.E., P.Eng., is Founder of Long International, Inc., one of the world's largest construction claims consulting companies, which also provides project management consulting services. Based in Colorado with offices throughout the U.S. and the Middle East, Long International focuses its practice on owners, engineering and construction firms, and contractors performing oil and gas, petroleum refining, petrochemical, chemical, power, LNG, mineral processing, manufacturing, industrial, building, and infrastructure projects worldwide. Mr. Long has over 50 years of U.S. and international engineering, construction, and project management consulting experience involving construction contract disputes analysis and resolution, arbitration and litigation support and expert testimony, project management, engineering and construction management, cost and schedule control, and process engineering. As an internationally recognized expert in the analysis and resolution of complex construction disputes for over 35 years, Mr. Long has provided construction claims consulting services and, in some cases, served as the lead expert on over 500 projects having claims ranging in size from US$100,000 to over US$2 billion. He has presented and published numerous articles and training seminars on the subjects of engineering and construction claims analysis, entitlement issues, CPM schedule and damages analyses, loss of productivity cumulative impact claims, and construction claims prevention.

Before forming Long International, Mr. Long was Senior Vice President, Contract Administration, for a major electrical and mechanical contractor. In this role, he had corporate-wide responsibility for technical management and oversight of the preparation and resolution of construction claims. In addition, he was responsible for the development, training, and implementation of project management policies and procedures to ensure that profit, cost, schedule, scope, quality, and safety objectives were achieved. For thirteen years, Mr. Long managed the construction claims practices of two large consulting firms. Prior to his consulting career, Mr. Long gained 13 years of project management and process engineering experience on petroleum refining, oil shale, synfuels, mining, and power generation projects with Tosco, Fluor, and Conoco.

Mr. Long earned a B.S. in Chemical Engineering from the University of Pittsburgh in 1970 and an M.S. in Chemical and Petroleum Refining Engineering from the Colorado School of Mines in 1974. He is a Registered Professional Engineer in the state of Colorado, and a P.Eng. in Ontario, Newfoundland and Labrador, Canada. Mr. Long is based in Littleton, Colorado, and can be contacted at rlong@long-intl.com and (303) 972-2443.

Douglas J. Nutter, Long International's Manager of Graphic Services, prepared the graphic illustrations throughout this book, as well as formatting the text for design and indexing. He has over 45 years of experience in design, illustration, cartography, and graphic production for technical support, arbitration, and litigation. Mr. Nutter is skilled in developing complex technical issues and data into concise presentations for use by counsel, juries, and arbitration panels. Previously, Mr. Nutter gained 12 years of engineering drafting, illustration and graphic design experience on petroleum refining, oil shale, synfuels, mining, and power generation projects with Tosco Corporation. Mr. Nutter was also Manager of Graphic Services for Kellogg Corporation in Littleton, Colorado for twelve years prior to joining Long International.

Mr. Nutter currently provides graphic design, illustration, and technical support for construction claim preparation, analyses, defense, and negotiation of settlements for various parties, including owners, contractors, transit agencies, universities, sureties, financial institutions, law firms, and architectural firms. Mr. Nutter is based in Littleton, Colorado, and can be contacted at dnutter@long-intl.com.

Claire E. Crevey, Long International's Assistant Graphics Manager, provided technical editing and proofreading for this book. She has more than 16 years of experience in professional graphic design, writing, editing, and proofreading. Ms. Crevey has worked with many diverse clients with unique needs, including consulting firms, small businesses, nonprofit organizations, and the publishing industry. She has designed, edited, and proofread complex documents including books, case statements, trade catalogs, and multi-page brochures. Her experience also includes branding, designing logos, business cards, web banners, brochures, infographics, illustrations, and websites. Ms. Crevey is based in Littleton, Colorado, and can be contacted at ccrevey@long-intl.com.

Acknowledgments

The author makes the following acknowledgments regarding certain of the material that has been included in this book:

Chapter 3, "Why Construction Disputes Occur and How to Promote Early and Amicable Settlements," is adapted from an article by **Michael Harris.** Mr. Harris is a Principal with Long International and has over 35 years of construction project and claims management experience. His comprehensive consulting and management background includes risk evaluation and managing and resolving complex claims and contract disputes on major construction projects throughout the U.S. and globally for a wide range of industries, including the power, water, petrochemical, industrial, commercial, healthcare, and general building sectors. Mr. Harris has extensive expertise in key construction disciplines including project design, project management, construction management, contract management, risk and claims analysis, on-site risk management, claims resolution, claims avoidance management, cost management, and expert witness report preparation and testimony. As well as his extensive project execution background, he also has considerable experience in various types of dispute resolution forums such as arbitration, litigation, and mediation proceedings in North America, Europe, the Middle East, and the Far East. Mr. Harris has also been appointed as a mediator and arbitrator on various international projects. He has been responsible for the management of numerous major projects, both from a construction and dispute perspective, ranging in size from US$100,000 to over US$500 million.

Chapter 4, "Assessment of Problems Associated with Poor Project Management Performance," is adapted from an article by **Truman D. King, P.E.** Mr. King was a Principal with Long International and has over 50 years of experience in engineering, procurement, and construction operations in the mining, oil and gas, refining, and real estate development industries. He has successfully completed assignments in the U.S., Middle East, and Europe. Mr. King was a Senior Project Manager for Worley Parsons, where he managed the Front-End Engineering Design and acquisition of Process Design Technology for a US$500 million expansion of an oil refinery in the UAE to produce clean diesel fuel. He has also managed oil and gas, refining, power distribution, marine and other engineering projects and provided analysis of and resolution of contractor's claims. Mr. King has extensive knowledge of the owner's and contractor's project management responsibilities on EPC turnkey projects. He has successfully analyzed claims involving mismanagement, defective engineering, delay, disruption, and disputed change orders.

Chapter 7, "Avoiding Project Failure by Defining Requirements," is adapted from an article by **Douglas A. Bassett, P.Eng., PMP, FCIP.** Mr. Bassett was a Principal with Long International and had over 35 years of Canadian and international consulting experience involving construction project management, project leadership and project governance/review. Mr. Bassett was a seasoned project leader. During his career, Mr. Bassett was involved in all types of oil and gas energy projects with a primary focus on oil sands and refinery facilities, offshore fixed and floating drilling and production facilities, and conventional onshore production facilities. Mr. Bassett's accomplishments included serving as project leader on Canada's first offshore development production platform, project leader in a successful alliance to develop a major oil sands mine, and governance leader of project approval reviews for oil sands and offshore mega projects. Mr. Bassett passed away in September 2020.

Chapter 11, Cost Control for the Project Manager, is adapted from an article by **Philip R. Moncrief, P.E.** was a Senior Principal with Long International, and has over 50 years of U.S. and international consulting experience involving engineering and construction, contract disputes, project management, project development, engineering/construction management, and process engineering. As an internationally recognized expert in the engineering and construction industry, he has managed large projects and had profit and loss responsibility for the engineering and construction operations of four major engineering-construction companies.

Chapter 12, "Effective Change Order Management," is adapted from an article by **James M. Bolin.** Mr. Bolin was a Senior Executive Consultant for Long International and has over 35 years of experience in all aspects of project and construction management, preconstruction planning, and project cost/schedule controls. His qualifications include claims and change order analysis and resolution, preparation of fixed-cost proposals and budget baseline planning, site management and coordination, prime and subcontract development and administration, materials expediting and procurement, and critical path method (CPM) delay analysis, schedule development, progressing and reporting. This diverse expertise has been obtained through involvement on numerous project types, including oil production facilities, pipelines, pump stations, water and wastewater treatment plants, mineral processing facilities, water storage, hydroelectric plants, hospital/medical, transit, airports, communications, and environmental remediation.

Chapter 13, "Keys to Successful Major Project Execution," is adapted from an article by **S. Andrew McIntosh.** Mr. McIntosh is a Senior Principal with Long International and has nearly 50 years of engineering, construction, energy and management consulting experience. He has extensive international greenfield and brownfield project development and project management experience in diverse areas such as oil and gas, including refinery, pipelines and offshore projects, LNG, petrochemical and ferrous metallics projects ranging in value from US$20 million to US$1.8 billion. He has provided expert analysis of technical issues and their causes and effects related to project management performance on multibillion-dollar EPCA projects. Mr. McIntosh also has extensive experience in the development of project management and project execution plans and procedures, stage-gate processes for project development, and the analysis and resolution of engineering and construction claims involving disputed change orders, schedule delay, and loss of productivity claims. He also completed expert analysis of technical issues and the causes and effects of an EPC contractor's project management performance on a major LNG project, and has prepared and submitted various independent expert reports. Mr. McIntosh has also provided expert advice to foreign governments on major LNG project development and natural gas monetizing strategies from natural gas offshore fields as well as from associated gas from offshore oil producing fields.

Chapter 14, "Project Team Motivators and Demotivators," is adapted from a chapter from *On Time and Under Budget: The Ultimate Power of Team Leadership,* a book written by **Michael J. Vallez, P.E., M.B.A., Lean Six Sigma,** published in 2014. Mr. Vallez is a Senior Principal with Long International and has over 40 years of hands-on and leadership experience in project management, engineering/construction management, cost and schedule control, change management, claims, dispute resolution, and mine and process engineering. He has served in executive management roles in industry, including both the owner and contractor sides with companies and contractors working on world-class projects for oil and gas companies, power companies, international mining companies, and other institutions. He has a proven ability to organize and integrate the work of multi-disciplined technical specialists and project construction teams to achieve corporate financial goals and objectives of ROI, safety, operational performance, cost, and time. In all, he has provided leadership on several billion dollars' worth of projects in the mining, power, oil and gas, industrial, heavy civil and commercial sectors. Mr. Vallez has written several books about construction management, safety, and effective project leadership.

Chapter 15, "Claims Can Be Mitigated If the Owner Coordinates and Cooperates on Multi-Prime Construction Projects," is adapted from an article by **Ronald J. Rider, M.B.A.** Mr. Rider is a Senior Principal with Long International and has over 30 years of comprehensive construction experience and expertise from over 100 projects in the areas of project management and dispute resolution services. He develops and monitors critical path method schedules, and provides consulting services that include schedule delay analysis, project estimating, project cost control, project billings and payments, change order pricing and resolution, contract administration, schedule recovery, project close-out, and dispute resolution of project problems. Mr. Rider is highly skilled in cost and labor hour variance modeling, impact identification and analysis, retrospective CPM schedule delay preparation and evaluation, concurrent delay assessments, project acceleration analysis, and damages verification and quantification.

Chapter 16, "Commercial Awareness Training for EPC Projects," is adapted from an article by **Roger D. Javid, M.B.A.** Mr. Javid was a Senior Executive Consultant with Long International and has over thirty years of experience on large international projects ranging in size from US$100,000 to over US$1 billion. Mr. Javid is an expert in project management, commercial management, development of project budgets, analysis and negotiation of EPC variation orders and amendments, management of project purchasing and contracting cycles, field supervision, project close-out, insurance claims, commercial awareness training, and construction claims preparation and settlement negotiations. He also has extensive experience with FIDIC contracts, Federal Acquisition Regulations, and USAID Acquisition Regulations. His project experience includes NGL/GOSP facilities, LNG facilities, offshore platforms and pipelines, petroleum refineries, petrochemical facilities, telecommunication fiber loops, wastewater treatment facilities, power generation facilities, and theme parks and resorts.

Chapter 19, "Construction Claims Analysis Checklist," is adapted from an article coauthored by **Richard J. Long, P.E., P. Eng.,** and **Rod C. Carter, CCP, PSP.** Mr. Carter is President of Long International and has over 20 years of experience in construction project controls, contract disputes, and resolution, negotiations, mediation/arbitration support, and expert testimony on scheduling, loss of productivity, and quantum issues. He has experience in entitlement, schedule, and damages analyses on over 30 construction disputes ranging in value from US$100,000 to US$7 billion, related to oil and gas, heavy civil, nuclear, environmental, chemical, power, industrial, commercial, and residential construction projects. Mr. Carter is proficient in the use of Primavera P6 and P3 software, and he has extensive experience in assessing the impact to engineering and construction works of RFIs, change orders, and other events. He specializes in loss of productivity, cumulative impact, and quantum calculations, and has had a lead role in assessing damages on more than a dozen major disputes. In addition, Mr. Carter has developed cost and schedule risk analysis models using Monte Carlo simulations to address the uncertainty of estimates and claims. He has testified as an expert in construction scheduling and damages, and he has presented expert findings to an international arbitral tribunal.

Why Construction Claims Occur and How to Prevent Them

Richard J. Long, P.E., P.Eng.

LONG INTERNATIONAL

Chapter 1

Setting the Stage

Chapter 1

Setting the Stage

Construction projects have invented their own version of opposites: the contractor and the owner. At some early stage in the game, often before the contract is seven days old, the two sides have identified their lack of common interest. Battle lines are drawn, and the war begins. Everybody knows what the war is about; the contractor wants to maximize profit and the owner wants to minimize cost. Rules of warfare are drawn up and are called contract terms and conditions and specifications. Attorneys, accountants, claims consultants, arbitrators, judges, and juries are employed to identify the winner. In scholarly circles, this form of warfare is identified as the adversary relationship. This clearly states that the relationship is not one of trust.

In practical terms, the project atmosphere has gone bad. The contractor's key personnel spend most of their time fighting with the owner's project representatives. To ensure the widest possible dispersion of this bad atmosphere, the owner's inspectors fight with the contractor's foremen. With all of these problems occurring, management of the project goes by default to the hourly worker.

The high frequency of contract disputes must be recognized, and all reasonable measures undertaken by each party involved in today's engineering and construction projects. Technical complexities, the increasingly contentious nature of the contract environment, tight competitive bids, and owners with tough-minded and sometimes inequitable contractual attitudes have caused the final act of many construction projects to occur in an arbitration hearing or in the courtroom. Legal and expert costs for preparing and defending claims have exacerbated an already high-risk business environment for all construction contracting parties.

Construction has always been a risky business; however, today's construction marketplace contains risk elements that go beyond the performance of field operations. These risks relate to outside factors that impinge on the performance of the field work, and thus cause substantial increase in the time and cost of contract performance.

Contract documents generate adversarial roles and inhibit the development of teamwork; examples are presented below:

- Project cost and schedule control systems are often not sufficiently sophisticated or integrated.
- Contract administration often does not recognize the expressed and implied terms of the contract.

- Regulatory control is complex and produces unfavorable economic conditions.
- Environmental considerations can produce severe time delays to the commencement and completion of the construction phase of projects.
- Building codes tend to be out-of-date, inapplicable and not uniform.
- Complex union agreements inhibit resource allocation and optimization.
- Design engineers untrained in construction contract issues often masquerade as construction managers.
- The legal process focuses on blame and inhibits problem solving.
- Sociological and political factors can create major project delays.
- Project management systems and procedures are often not state-of-the-art practices.

Although a construction project should be a mutually beneficial endeavor, adversarial relationships that can culminate in disputes are common. Owners and contractors are acutely aware of the significant monetary, labor, and equipment resources that they must commit to a construction project and the adverse effect any increase in these resource commitments would have on their bottom lines. Joint venturing or partnering have been suggested by many industry leaders to lessen this natural economic tension and alleviate disputes. Regardless of the merits of these proposals, the reality of the construction industry is that contractors and owners must be aware of and educated about claims and dispute resolution to protect their significant resource investment.

Many contractors no longer enjoy a trusting relationship that brings repetitive contracts with the same owners. Owners have become more global in their operations and must establish relationships with contractors outside their geographic comfort zone. A handshake meant everything in the distant past. Contract language seemed subordinate to the course of dealings that developed over many years of interaction between owners, engineers, contractors, and subcontractors. The "handshake" atmosphere has been replaced with a very competitive bid arrangement that usually results in an owner meeting a different contractor on each new project. Scar tissue that was developed from contentious issues on previous projects often finds its way into tougher terms and conditions. Because the parties are strangers, an adversarial relationship is likely to develop when problems arise.

Parties are enthusiastic and optimistic at the inception of every construction project. They are confident of their capability and commitment to fulfill their respective obligations under the contract documents while they mutually benefit from the project. Unfortunately, this enthusiasm and optimism are often dashed very early in the project and the "honeymoon" is quickly over. Immediately, the contract documents are reviewed with a fervor and intensity previously considered unnecessary and unwarranted. Yet these contract documents define the primary relationship between or among the parties and form the basis, in general, for all claims.

The possibility of being blindsided by a large claim has caused owners to question basic construction contract terms and conditions and the way they contract for design and build. Owners create exculpatory contract clauses to protect against the reoccurrence of such claims. When their workload is down, contractors take risks by accepting tough contract terms and conditions.

Owners expect more from their engineers, whether in-house or outside design professionals, but insist on lower costs and time to produce "issued for construction" documents. The engineering profession is continuously examining the quality of its product to determine if errors and omissions are increasing because of regression in professionalism or skill, or because of inadequate budgets and time for design work. If errors and omissions are increasing, how can they be better controlled? If they are not increasing, how can they be further reduced?

Professionals who prepare contracts and specifications are subjecting their work to continuous revisions to see if simplifications, exculpatory language, or new concepts are needed. Many defensive contract clauses such as "no damage for delay" and the usual subsurface exculpatory requirements to examine the site and not rely on borings have often failed as a result of improper contract administration or inadequate site investigations by the owner. Specifications are too often a "cut and paste" exercise that does not meet the varying needs of the project. Numerous claims could be avoided by knowledgeable and careful specification writing incorporating knowledge about construction law and claims.

1.1 Characteristics of a Highly Impacted Project

Owners, engineers, construction managers (CM), and contractors have the best of intentions when they initiate a project. The owner wants a profitable, state-of-the-art facility, the engineer desires to create the owner's intent in its design, the CM agrees to manage the quality, safety, cost and schedule objectives of the owner, and the contractor wants to build the facility on time and make a reasonable profit. Then the parties sign a contract, and everything changes.

Too many projects are transformed from the best of intentions into highly impacted projects that are out of control and headed for claims and disputes. The following characteristics of these projects are all too common:

- A fast-track concept is employed to reduce the time and costs for completing the project.
- The contractor bids on incomplete design documents.

- Many addenda are issued during bid preparation.
- The contract is awarded to the lowest bid contractor but not necessarily the most qualified.
- The contractor issues an inordinately high number of requests for information (RFIs).
- The owner/engineer issues a blizzard of changes.
- The owner (or construction manager (CM)) is inadequately staffed to process the multitude of change orders and respond in a timely manner to the numerous RFIs from the contractor.
- The contractor fails to prepare an adequate as-planned Critical Path Method (CPM) schedule to identify the planned sequence and duration of work activities.
- The contractor's cost and schedule control systems are not integrated.
- The contractor does not employ a system to capture the actual cost of changed or impacted work.
- The contractor fails to maintain an accurate and updated progress schedule including identification of changes and causes of delays.
- The contractor requests time extensions that are denied because of inadequate support, failure by the owner to recognize excusable delays, or inflexibility in the project completion date because of regulatory compliance or other constraints.
- The owner employs language in the change orders stating that the price adjustment includes all direct, indirect and impact costs.
- The contractor signs the first few change orders with the owner's language, but later refuses to sign off its rights to claim impact costs because of the blizzard of changes and delays.
- The project is completed late, the owner has paid a significantly higher price for many changes, and the contractor now has a big delay and impact claim, and the owner withholds payments to cover its liquidated damages (LDs).

How are the owner's and contractor's damages on a highly impacted project best resolved? The best solution is to avoid the sins of poor project management and contract administration identified above. However, we do not yet live in a perfect world. Until we do, we will continue to face these problems—all too often on a retrospective basis.

1.2 Effect on Project Costs

Large value claims, when asserted successfully, may destroy the viability of a project or make construction more difficult by adding to the disenchantment between the owner and contractor. If a sizable claim is found to be valid, an owner faces serious problems. New financing may be required, later occupancy may be necessary, income from operation of the facility or rent from commercial property may be delayed or lost, and staffs may have to be funded longer than planned to administer the project.

Most owners include contingency funds in their project budgets to cover the cost of claims. These contingencies are normally established at the start of a project to preclude the need for special funding actions later. The contingency funds, however, are usually limited and seldom are large enough to handle claims, which can equal one-third or more of the base cost of the project. In a recent refinery upgrade project having a $42 million contract value, the engineering, procurement, and construction (EPC) contractor filed a $45 million claim. Resolution of this claim took over three years and the expenditure of nearly $10 million in combined legal and expert costs by the combatants.

If a project experiences unknown and unfunded cost increases, emergency actions may be required, such as a temporary halt to construction, robbing unspent funds from other budgeted projects, requesting special appropriations, financing by a new tax or bond issue, or additional borrowing.

Almost any of these solutions create serious problems for the owner, as well as the contractor, because lost time costs money. Nonreceipt of payments resulting from incurred costs not envisioned in the contract price may result in the contractor's financial capability becoming dangerously unstable, its being unable to pay its subcontractors or suppliers, and loss of bonding capacity. Subcontractors, in turn, may default, and the project may collapse due to a lack of resources and money.

1.3 The Court's View of Contract Interpretation

In view of the claim-causing problems discussed earlier, the need for a contractor to file a lawsuit may be for its very survival. Over the years, many landmark cases have evolved new avenues and bases for contractor recovery, some of which are as follows:

> *But if the contractor is bound to build according to plans and specifications prepared by the owner, the contractor will not be responsible for the consequences of defects in the plans and specifications. This responsibility of the owner is not overcome by the usual clauses requiring*

> *builders to visit the site, to check the plans, and to inform themselves of the requirements of the work....*[1]
>
> *... It is, however, an implied provision of every contract, whether it be one between individuals or between an individual and the government, that neither party to the contract will do anything to prevent performance thereof by the other party, or that will hinder or delay him in its performance.*[2]
>
> *Contractors are businessmen usually pressed for time and consciously seeking to underbid competitors. Consequently, they estimate only those costs that they feel the contract terms will permit the government to insist upon in the way of performance. They are not expected to ferret out hidden ambiguities or errors in the bid documents and are protected if they innocently construe in their own favor an ambiguity equally susceptible to another construction or overlook an error.*[3]
>
> *Merely because the highway and bridge construction contract was complete and the job accepted well before the contract completion date did not preclude the contractor's recovery for increased costs due to delay where, absent delays caused by the state, the contract could have been completed earlier and substantial sums of money saved by contractor.*[4]
>
> *Ascertainment of damages is not an exact science and it is not essential that the amount of damages be ascertainable with absolute exactness or mathematical precision; it is enough if the evidence adduced is sufficient to enable a court or jury to make a fair and reasonable approximation.*[5]
>
> *Contracting authority has the duty to invoke its contractual rights to compel cooperation among contractors. Where the transit authority limited the exercise of its supervisory authority to arranging an unsuccessful conference between adjoining contractors to resolve the problem of water overflowing from the job site of the first contractor onto the job site of the second contractor when it had a duty to compel cooperation, resulted in a break of contractual duty owed to the second contractor.*[6]

While the owner has the most at risk in terms of money and in terms of the urgency of prompt and on-time completion of the project, he also has—from the beginning—the greatest opportunity to control how the project is conducted and to provide appropriate organization and methods for preventing disputes from arising and for solving disputes promptly as they do arise to minimize their effect on the project as a whole. This opportunity cannot be emphasized too strongly. If the owner pays careful attention to this, there is no question but that he stands to have a more efficient and, therefore, less

1 *United States v. Spearin*, 248 U.S. 132 (1918).

2 *George A. Fuller Co. v. United States*, 69 F. Supp. 409 (108 Ct. Cl. 1947).

3 *Bromley Contracting Company*, ASBCA 14884, 16045 72-1 BCA ¶ 9252 (1972).

4 *Grow Constr. Co. v. State*, 391 N.Y.S.2d 726, 729 (App. Div. 1977).

5 *Metropolitan Sewerage Comm'n v. R.W. Constr.*, 78 Wis.2d 451, 255 N.W.2d 293–302 (1977).

6 *Shea-S&M Ball v. Massman-Kiewit-Early*, 606 F.2d 1245 (D.C. Cir. 1979).

costly project and that he will substantially diminish the risk of costly litigation or arbitration after the project is completed.

As one judge has commented:

> *Except in the middle of a battlefield, nowhere must men coordinate the movement of other men and all materials in the midst of such chaos and with such limited certainty of present facts and future occurrences as in a ...construction project.... Even the most painstaking planning merely turns out to be mere conjecture and accommodation of changes must necessarily be of the rough, quick and ad hoc sort, analogous to ever-changing commands on the battlefield.*[7]

Therefore, the object for the owner will be—from the very beginning of the project—to eliminate, as much as possible, the conjecture and the need to accommodate change on an ad hoc basis, to make sure that all the various elements in the project are coordinated to the best possible extent, and to be sure that when he needs to make decisions during the project, he is able to make them with a full understanding of all the facts and consequences. This can best be done using carefully planned and prepared contract documents, plans, and specifications.

For some projects such as a major facilities expansion program, unique and complex design, scheduling, and construction requirements above a level considered routine may require special attention to the procurement and contracting process. It is at the beginning of the large project, before the designer is employed, that the owner can best control procurement procedures and contracts and minimize the potential for claims.

---

7 *Blake Construction Co. v. J. C. Coakley Co.*, 431 A.2d 569,575 (App. D.C. 1981).

Chapter 2

Situations Causing Costly Claims

Chapter 2 – Situations Causing Costly Claims

Chapter 2

Situations Causing Costly Claims

In a perfect world, the owner never changes his mind, the engineer never alters his drawings, the contractor never malperforms, the resident engineer's decisions are perfect, and Mother Nature behaves herself. That perfect world does not exist.

In the real world, with geometric precision, the forces of owner, engineer, contractor, and Mother Nature combine to make change. The owner changes his mind. The engineer changes his drawings. The contractor fails to manage his job. Mother Nature then changes what the others have missed.

The contractor who proudly boasts, "I've never had a claim," is a vanishing breed. To ensure survival in today's construction market, the contractor has learned to understand his rights and the techniques of survival when a claim arises. The owner too must ensure profitability of his project by defending unsupportable claims by a "lowball" contractor who intends to recover his bid errors by change order requests or end-of-project impact claims.

Depending on which party is assessing and determining the causes for claim generation, the perspectives cover a broad range. Voluminous claims are not foolproof indications that a compensable claim exists. It is frequently necessary to examine the project history in detail before intelligent judgments can be made on the validity of the claim.

When claim experience is evaluated to identify features of the contract, performance of the work, contract administration and management processes, and disputes resolution procedures that are associated with claims and, therefore, may be of critical importance in claim prevention, the following section discuss the major factors:

2.1 Contractor Practices that Cause Claims

2.1.1 Inadequate Site Investigation Before Bidding

Contractors often venture into new geographic areas to broaden their opportunities. Too often, the contractor will bid work without performing an adequate site investigation into areas such as availability of qualified labor, labor productivity, weather patterns, underground conditions, and other competing projects in the area that would compete for the available labor. As a result of not doing their homework adequately, contractors may then take every opportunity to prepare change order requests or file claims to make up for their bid errors.

2.1.2 Unbalanced Bidding Resulting from Manipulating Unit Prices and Front-End Loading Progress Payment Weighting Factors

Contractors often gamble by proposing high unit prices on types of work where they anticipate significant growth from the owner's bid quantities and lower prices on types of work where they feel that they need to be competitive. If these tactics prove to be ill advised, or if the contractor guesses incorrectly on which units will grow, the contractor may not be recovering its actual costs. In addition, contractors will try to get ahead on the cash flow curve by putting higher progress payment weighting factors (than could be justified on a balanced cost basis) on work performed early in the project to the detriment of work performed later in the project. Another approach may be submitting a higher percentage completion estimate in its pay requests for easier work, such as installing straight run pipe, and then experiencing lower productivity and higher unit costs for valve stations or small-bore pipe installations that are more difficult to install.

2.1.3 Bidding Below Costs and Overoptimism

This practice would be considered a deliberate underbid, presumably justified to break into a new geographic market or new line of work. After the euphoria of winning the job disappears and the reality of losing big money on the project sinks in, contractors may be persuaded to seek every opportunity to submit requests for change orders or file claims.

2.1.4 Poor Planning and Use of Wrong Equipment

The means and methods contemplated by the contractor's estimators to achieve the productivities anticipated by their bids are often not implemented in the field because inadequate planning is performed prior to mobilizing to the site. When lower than anticipated productivities are experienced, contractors will be on extra alert to identify claim opportunities.

2.1.5 Inadequate Cost and Schedule Control Systems

These problems are associated with imperfect management planning and control. Possible explanations for why estimated costs for project activities are so inaccurate when compared to actual costs, and why planned schedules experience significant delay, include lack of integrated costs and schedule control systems, poor definition of work activities, improper control of schedules and cost segregation, and inaccurate cost allocation methods.

2.1.6 Failure to Follow Authorized Procedures

The owner's specifications often require more expensive and time-consuming procedures for performing the work than recognized by the contractor when preparing its bid. When the owner rejects the contractor's work method and insists on the specified procedures, the contractor may attempt to file a claim.

2.1.7 Refusal to Proceed with Directed Work Pending a Contract Modification Despite Contract Clauses Giving the Owner the Right to Make a Unilateral Change

Owners often disagree with the contractor's price for performing changed work, but still want the work performed. To overcome this problem, contract provisions are often included to give the owner the right to make a unilateral change. If the contractor is experiencing cash flow problems or potential delays in completing the work, it may choose to play hardball with respect to forcing the owner to formalize the change and agree to its price and time extension request before performing the changed work. When the work is eventually performed, the actual cost may be much higher than it would have been had the work been performed when the change was identified. These situations are ripe for claims.

2.1.8 Performing Defective Work

Owners may need to withhold progress payments from contractors that have performed defective work. The delay to the project completion caused by the time required to correct these defects may also justify the owner's liquidated damages claims.

2.1.9 Slow Mobilization

Many contractors do not perform sufficient planning in advance to effectively mobilize for the project. After contract award, the owner informs the contractor that the site is available and ready for mobilization to begin. In a design-build or EPC contract, the start of construction is often optimistically planned to occur earlier than when the approved for construction drawings are completed, and equipment and materials are procured and delivered. Also, contractors may not move on to a site in an efficient or timely manner because of insufficient planning regarding the equipment and labor needs that are required.

2.1.10 Failure to Allocate Resources

Earning a profit is essential to any contractor's strategy. A contractor normally makes a profit by managing a project with the least amount of people and equipment on site and completing the project as quickly as possible. The more men and equipment that are situated on site, the more cash outflow is required. Managing limited amounts of manpower and equipment efficiently and cost-effectively is easier said than done, especially if a contractor is dealing with several projects at once.

What frequently happens is that a contractor either does not have the right mix of people or lacks the right equipment on a particular project. This happens because these limited resources are being utilized on other projects. Labor and equipment availability are important considerations to consider when preparing to bid on various types of construction projects. The project starts off slower than the early start dates, and acceleration and higher peak labor requirements are later needed to make up for these early resource availability delays.

2.1.11 Brokering the Work and Passing All Problems to the Subcontractors without Adequate Coordination

General contractors or large EPC contractors often subcontract all work but fail to adequately schedule and coordinate the work of the various trades. The general contractor or EPC contractor then points to the clause that it inserts in its subcontracts stating that the subcontractor must coordinate and cooperate with the other trade subcontractors. These problems often result in claims by the subcontractors who are impacted by the main contractor's failure to perform its coordination duties.

2.2 Owner Practices that Cause Claims

2.2.1 Competitive Bidding

Competitive bidding is the procurement practice in which the contractor who leaves the most money out of his estimate, usually as contingencies, gets the job. The lowest bidder is also the one faced with the highest danger of loss if contingencies, *i.e.*, risks, are encountered. Only the very naive owner is unaware of how much time is spent by the low-bid contractor in search of ways to cut losses by seeking additional funds through changes. While competitive bidding is often a sound procurement practice, the owner must recognize that a bid that is significantly below the other bidders is a claim waiting to happen.

2.2.2 Fixed-Priced or Lump-Sum Contracts

Each adjustment to the time or price of the contract requires a turning of the administrative machinery. All too often it becomes a contest; the contractor looks for ways to get more and the owner looks for means to give less. The contest becomes confrontational, and the dispute is formed. Allegations of scope change, and defective and deficient contract drawings and specifications are at the core of claims on these types of contracts.

2.2.3 Unit-Price Contracts

Owners often go out for bid seeking unit prices for performance of an incompletely defined scope of work. Justification for this pricing strategy is to reduce the overall execution time for the project, *i.e.*, a fast-track schedule. However, disputes often erupt as the engineered quantities are defined during the detailed design phase, and these engineered quantities are significantly more or less than the owner originally estimated. Claims for delay and the loss of productivity costs of significantly higher levels of required manpower that result from higher quantities than originally estimated, or under-absorbed overhead for lower quantities than originally estimated, are encountered when significant discrepancies occur between the original and final quantities.

2.2.4 Contract Language that Inappropriately Shifts Risks to the Contractor

The ideal contract is a balanced document that respects the rights of both parties and thoughtfully allocates each category of risk to the party who is in the best position to manage that risk. In the real world, the owner-drafted contract tends to lean sharply in the owner's favor using exculpatory language. While it may give the owner some degree of added protection, it also leads to higher bid prices and expensive arbitration hearing or courtroom battles.

2.2.5 Failure to Grant Time Extensions for Excusable Delays

Once upon a time, folks did not fuss much about time extensions. When the contractor finished 100 days late, the owner granted a 100-day extension—a simple transaction. It worked because contractors thought "claims" was a dirty word and owners did not have to use liquidated damages as a weapon. Today, most contractors think of "claims" as a proper noun. They also know the definition of constructive acceleration: the owner's denial or failure to act on an allowable request for more time. Today, most owners have learned—some the hard way—that requests for time extension should not be stored in the filing cabinet. Each request demands analysis and timely response.

2.2.6 Major Changes in the Plans and Specifications during Construction

The owner is ultimately responsible for defining the scope of a project during the conceptual phase of the project; the scope will probably change quite frequently without causing any undue hardship. However, changes that occur during the construction phase or even detailed design phase inevitably make life difficult for the construction contractor. Scope changes result from many possible situations, such as owner preferences, design reviews, constructability issues, safety requirements, hazardous operations reviews, and operations and maintenance concerns.

2.2.7 Mishandling of Minor Change Orders

The tranquility that follows contract award does not last long before the inevitable changes begin: the owner wants alloy pipe instead of carbon steel because chlorides are found in the feedstock; the engineer wants to move vessels higher in the structure to obtain more head on the pumps below because of higher pressure drop in the system; the contractor wants to use a less costly concrete batch plant instead of buying concrete from the local supplier, but the permits did not provide for a batch plant; and Mother Nature brings a 200-year flood. Change is the necessary ingredient. Without it, disputes are rare events. To minimize the risk of dispute, the owner should administer changes with three objectives in mind. First, try to obtain an up-front agreement (forward pricing) on time and cost. Second, try to include complete accord and satisfaction in the change order agreement. Third, do not let unresolved changes accumulate.

For example, the contractor wants to save a million dollars by building the concrete batch plant. The engineer blames the contractor because the local supplier wanted a higher price for delivered concrete than the contractor included in its bid. The contractor blames the owner for its reluctance to amend its permitting documents. The insolvable dispute ensues. What should the contractor and owner do? Reserve courtroom space and head for high ground? Hardly. Complex disputes can be untangled and resolved, but it demands prompt and decisive actions by the owner and the contractor.

Several guidelines are recommended. First, ignoring the problem will not make it go away. Second, each involved party should provide detailed documentation with respect to the basis of their positions. Third, try to ignore natural biases as the facts are examined. Owners lean instinctively toward their own team, but the leaning is not helpful when the contractor has the winning argument. Fourth, look with vigor for ways to mitigate damages. Fifth, be prepared to compromise as new facts emerge. The owner may not be persuaded to file an amended permit until the contractor shows that the dust emissions from the batch plant will be less than the dust created by the concrete truck traffic. Sixth, do not let unresolved disputes accumulate. Their gathering weight can have a deadly effect: they absorb staff time, poison relationships, drain the contractor's cash reserves, and steadily harden until they are an unresolvable mass.

2.2.8 Unrealistic Schedules and Underestimated Costs

Whenever time schedules and/or project costs are based on optimistic predictions of labor productivity, availability of skilled labor, good weather, and quantities based on an incomplete design, etc., the project will be burdened with missed milestone dates, cost overruns, and even low morale among project personnel. Sufficient time must be allocated in the planning stage of a project to accurately estimate time and cost requirements.

2.2.9 Failure to Obtain Adequate Funding

Adequate funding for construction projects is more difficult when money is tight. Financial institutions are often averse to adequately fund small to medium-size construction projects. Stringent guidelines are now enforced in the lending of money to construction owners. The banking industry is becoming much more selective in the kinds of projects in which it is willing to invest. The construction contractor has to be concerned about whether the owner is going to be able to reimburse or pay progress payments in a timely manner. By far the most common cause of contractor bankruptcy is not failure to earn a profit over the long term, but not being paid in a timely manner. Most contractors go bankrupt because they cannot meet short-term obligations, resulting from not being able to manage their current obligations.

2.2.10 Inadequate and Ambiguous Scope Definition in the Bid Documents

This has a far-reaching effect because a project's scope must be defined in enough detail for comprehensive scheduling, estimating, and resource allocation to be reasonably accomplished. Whenever the owner is unwilling to make a concerted effort to define the construction project in the early stages, or charges ahead using a fast-track concept without adequate contingency, problems will undoubtedly occur throughout the entire project life cycle.

2.2.11 Inadequate Time Provided for Bid Preparation

Owners are always under economic, market, environmental, political, and other pressures to start and complete a new capital project or a modification to an existing facility. As a result of these pressures, owners may hastily prepare incomplete bid packages and then request fixed-price or unit-price bids by a date that does not provide the contractor with an adequate time period to investigate the site conditions, thoroughly read and understand the owner's specifications, prepare its design, verify the accuracy of the owner's quantities, secure quotations for equipment and materials, prepare an adequate project schedule, and estimate its costs for performing the work. When this occurs, the contractor can either choose not to bid, add significant contingencies to its estimate to cover the risks and hope that its bid is still within the owner's budget, or bid competitively despite the risks and hope that if it wins the bid, it can recover any unanticipated costs through change orders and/or claims.

2.2.12 Excessively Narrow Interpretation of the Plans and Specifications

Plans and specifications are often susceptible to different but reasonable interpretations, particularly if they are unclear or ambiguous. To be competitive, contractors are always seeking reasonable ways to reduce their costs and win bids. If the contractor's interpretations are reasonable, and the owner rejects the contractor's equipment and/or material submittals or construction means or methods because its intent was different, claims may be the only recourse available to the contractor, particularly if the owner's interpretations are very costly.

2.2.13 Owner-Furnished Materials

Owners typically want to procure and furnish their own equipment for unique projects, such as chemical plants and refineries, educational facilities, hospitals, and research facilities, because of favorable purchasing agreements with vendors or the special needs associated with such projects. Most owners fail to understand that contractors have specific materials management plans for the project—a plan for when that material will be installed and where it is going to be stored once it arrives at the site. Most owners have the mistaken notion that whenever owner-supplied material arrives on the site, the contractor can efficiently store or install the materials. The sequence of delivery of bulk material items such as structural steel, large bore piping, and piping supports is extremely important to the contractor's efficiency. Owners and contractors must thoroughly coordinate and schedule the delivery of these items if the owner elects to perform the procurement function. Contractors usually have installation schedules that they plan to follow, and this type of misunderstanding leads to significant material management problems.

2.2.14 Incomplete and Defective Plans and Specifications

There have been few instances where an engineering design was so complete that a project could be built to the exact specifications and drawings as contained in the original design documents. There are always going to be errors or omissions in the design documents that will result in some modifications to the originally planned construction process.

2.2.15 Delays to Project Award

Often as a result of delays in securing funding approvals or permits to commence the contract work or design completion delays by the owner's engineers, owners may delay project awards beyond the dates stated in the bid documents. It is not uncommon, however, for owners, after clearing such roadblocks to their projects, to still insist that the contractor complete the project by the original bid completion schedule and at the same price. Contractors must then evaluate the tradeoff of lower time-related costs because of the shorter time frame to complete the project and the potentially increased costs because of higher manloading, overtime, shift work, or other acceleration efforts to complete the same scope in a shorter time frame. If the contractor prepared a resource-loaded CPM schedule during the preparation of its bid, it could evaluate the effect of such acceleration before agreeing to the same completion date despite the contract award delay. If the contractor did not prepare a CPM schedule during its bid preparation, it may have to take the risk that it could still meet the completion date. If the contractor accepts the project award at its original bid price, and its increased costs to complete the project on time significantly outweigh the savings in time-related costs, a contractor may submit a claim to recover its increased costs.

2.2.16 Failure to Give Adequate and Timely Access to the Work Site

When an owner does not provide the contractor with adequate access to the work site to complete its work in the time frame and in unobstructed conditions that were anticipated in its bid, the contractor may file a delay and loss of productivity claim for the resulting increased labor costs resulting from the access problems.

2.2.17 Interfering with Work on the Job Site

Owners invite claims by inappropriately directing the contractor's work, performing concurrent operations, or actively interfering with the contractor's least-cost performance.

2.2.18 Letting Other Contracts in the Same Area

A contractor depends on the availability of qualified craft labor in adequate numbers to perform the project work in the quality, time frame, and productivity rates anticipated by its bid. When the owner disrupts this labor supply by letting other projects in the area without prior notice of such plans at the time of the subject project bid, contractors may submit claims for increased costs of importing labor from outside the geographic proximity of the project, paying higher wages to attract more labor, paying overtime premiums to attract more labor, or incurring decreased productivity for less experienced workers.

2.2.19 Failure to Recognize Legitimate Change Order Requests

The owner's conduct during construction is a critical factor. Disputes will flourish if the owner elects to stonewall all requests for additional time and money. Even the most contentious of contractors will come forth with some well-founded requests that should be handled with equity and dispatch. The owner's positive response to reasonable requests can set the tone for the subsequent relationship of the parties.

Rather than flatly denying all requests, the best approach is a balanced one: pay the reasonable requests and save denials for the spurious. It is not a guaranteed solution, but it is the best policy that the owner can adopt.

2.2.20 Delaying Decisions

Late decisions that delay the performance of the contractor's work are invitations for claims. Examples of this problem are decisions on pending change orders that drag on well beyond the contractor's ability to complete the work efficiently and without delaying the project, and decisions to approve a time extension or require the contractor to accelerate the work.

2.2.21 Delayed Approval of Submissions, Shop Drawings or Materials

These types of delays can often affect activities on the critical path of the project schedule and cause delay claims. The contract should state the turnaround time that is available to the owner to perform these tasks. In addition, the contractor should always state in its transmittals the date when it needs a response.

2.2.22 Overzealous Inspections and Rejection of the Contractor's Work

Holding the contractor to higher than specified standards for performing the work can cause delay claims and claims for increased costs for correcting the work to the inspector's unreasonable standards.

2.2.23 Inadequate Project Management Controls

Project management control areas that cause the greatest concern include lack of meaningful cost and schedule control information because the contract does not require detailed cost and schedule information to be provided by the contractor; project decisions that are not made in a timely manner or not promptly implemented; confusion regarding the apparent authority and responsibility of project personnel; and, an apparent lack of integrated planning.

2.2.24 Refusal to Accept Materials or Equipment that Meet Specifications

This problem often occurs over the interpretation of "or-equal" specifications, where the contractor proposes equipment or materials that the contractor believes meets the owner's specifications and requirements, but the owner insists on a specific manufacturer or model that is usually more costly than the contractor's submittal. If the owner wants a specific item, he should specify the exact item and not include "or-equal" language in the specifications.

2.2.25 Bidding the Project Construction with an Incomplete Design

In its eagerness to get a construction contractor started in the field, theoretically to complete the project sooner, the owner will issue its bidding documents prior to the completion of "ready for construction" documents. Contractors usually only bid on what they see in the documents and are reluctant to add contingencies for unknowns because such bids may not be competitive. Completing the design often brings claims for changes and associated delays that, in the long term, extend the project duration and increase its costs.

2.2.26 Assignment of Inexperienced Staff to Oversee the Contractor's Performance

Many owners undertake projects with inexperienced and inadequate staff to appropriately administer the contract in a timely manner. Such problems often lead to delays waiting for the owner's approval of submittals, changes, or responses to RFIs. Alternatively, the owner could retain construction management, contract administration, or project controls personnel to increase its staff, but this option may not be implemented because of budget constraints. The result of this problem is usually delay, disruption, and claims by the contractor for increased costs.

2.2.27 Unanticipated Site Conditions

Owners often do not undertake adequate underground investigations to determine the type of soil, bedrock, existence and depth of water, contamination of soil, and other subsurface conditions to provide the contractor with adequate information to prepare its bid. The naive owner will put disclaimers in its bidding documents in an attempt to put the risk on the contractor. The results of finding these "differing site conditions" are delays and claims for increased costs.

2.2.28 Contract Requirements for Socioeconomic Objectives

Public or government owners will often place requirements in their contracts for achieving certain hiring objectives, such as minority percentages or local hire percentages. If the number and skill levels of the available labor are inadequate to meet these objectives, the project may be delayed, and the contractor may attempt to recover any increased costs it has incurred because of an "impossibility of performance" situation.

2.2.29 Unclear Definition of Completion of Work

Owners typically designate the Mechanical Completion Date or the Substantial Completion Date as the key completion milestone in the contract. When these dates have occurred, the owner can begin to start up the facility or begin to move in and partially occupy the building. Also, this date is usually the date from which liquidated damages are measured. Final Completion is the last milestone before the contractor can completely demobilize. Often, the contractor is merely completing punch list work or certain miscellaneous work that was not necessary for Mechanical Completion or Substantial Completion. Disputes often erupt at the end of the project if the owner's contract definition of Mechanical Completion or Substantial Completion is unclear. The contractor knows what it must do for Final Completion, but each owner's requirement for Mechanical Completion or Substantial Completion may be different. For example, are as-built drawings, hydrotest packages, operating manuals, and loop check documentation required for Mechanical Completion or Substantial Completion? If the owner requires these items before it will agree to the contractor's declaration of Mechanical Completion or Substantial Completion, but the contractor was planning to submit these after it declared Mechanical Completion or Substantial Completion, a dispute may occur over liquidated damages.

2.3 Causes Associated with the Contract Documents

2.3.1 Exculpatory Clauses

Such clauses are an attempt to shift the risks from the drafter of the contract to the other party. However, if one party cannot control such risks and, therefore, incurs increased costs, his only means of recovery may be a claim. A prudent contractor will include contingencies in its bid price for these risks. An example is a "no damage for delay" clause.

2.3.2 Mandatory Advance Notice of Claims

Owners, construction managers, or prime contractors will often insert notice clauses in their contracts requiring the other party to provide immediate notice of events giving rise to a claim. While these warning signs may be helpful in the mitigation of increased costs and time of performance, it may not always be possible to identify the existence of such problems within a few days of the potential claim-causing event. Even more difficult is the requirement to accurately price the costs associated with claim-causing events, particularly when impact ripples throughout the project. If forced to prematurely quantify the price for a change, the contractor will include high levels of contingency or impact costs in its change order prices. The owner's unfair rejection of changes because of high costs or lack of notice will certainly create a claim situation.

2.3.3 Finality of Field Engineer's Decisions without Review or with a Possible Conflict of Interest

The design engineer that is assigned to coordinate with the contractor's personnel often reviews requests for change orders. Legitimate changes are often denied because in accepting such changes, the designer may be admitting liability for errors and omissions. The innocent contractor may have no choice but to file a claim.

2.3.4 "Differing Site Conditions" Clauses

The intent of most differing site conditions clauses is to allow a contractor to recover its increased costs for "differing site conditions" not identified in the contract documents, and thereby allow the owner to receive lower bids without contingencies for unknowns. Inappropriately worded clauses, which attempt to put the burden on the contractor for such unknowns, will usually result in a claim in that the contract will seek fairness from the owner or the trier of fact if an amicable settlement is not achieved.

2.3.5 Lack of Periodic and Independent Review of Documents before the Bid

An independent review of the contract documents for ambiguities, conflicts, potential constructability problems, and design and specification errors is always a wise investment for an owner cautious about claims.

2.3.6 Use of Nonstandard and Conflicting Forms

An unsophisticated owner may attempt to assemble its contract documents, reporting forms, and other technical requirements without assuring that such documents are well coordinated, free from conflicts, and unambiguous. When claims result from such problems, the results are usually in favor of the party who did not prepare the documents.

2.3.7 Omissions of Clauses Needed for Effective Contract Management

Properly worded contract clauses defining the requirements for as-planned schedules and schedule updates, progress payments based on a balanced schedule of values and a procedure for measuring the completion percentage, handling of differing site conditions, adjustments for variations in quantities, change order pricing and direction to proceed with changes if no agreement is reached on price, acceleration for contractor-caused delays, ownership of float, analysis of delays, turnaround time for submittals and RFIs, and procedures for the resolution of disputes are vital to the proper management of a contract. Absent such clauses, the parties will often act in an inappropriate and costly manner, creating a project environment ripe for claims.

2.4 Causes Associated with Contract Awards

2.4.1 Diversity of Contract Award Rules

On public jobs, government agency rules may vary widely as to award of the contract in the event of bid problems such as:

- The low bidder may not meet the minority or local labor percentage requirements.
- The low bidder may have included qualifications or exceptions in its bid that are controversial.
- There may be conflicts of interest.
- Low bid contractors whose bids are rejected because of such conditions may file claims against owners, tying up the project for months.

2.4.2 Treatment of Bid Errors

Claims over bid mistakes are common. The contractor will claim that such arithmetic mistakes were not errors in judgment and should not allow the owner to gain an unfair price for the work.

2.5 Causes Associated with Contract Administration

2.5.1 Lack of Timely and Effective Management Response

Personnel charged with the responsibility of responding to the contractor's requests for change orders, time extensions, and legitimate requests for equitable adjustments to their contract amounts must perform their duties in a timely manner. A deaf ear to the contractor's repeated inquiries to the status of the processing of these issues can leave a contractor no choice but to take the issue to a higher level, file a claim, or file an arbitration demand or a lawsuit.

2.5.2 Lack of Coordination of Contracting Agency Responsibilities

When dealing with government agencies, contractors often become frustrated with a lack of coordination responsibility between the various departments involved in the project. An unresolved issue can become a hot potato between various agency departments, neither of which may want to deal with the problem. To get someone's attention, the contractor may have to file a claim, arbitration demand, or lawsuit. These alternatives could be avoided if clear responsibilities were identified in the contract documents and coordination procedures for the project.

2.5.3 Inadequate Delegation of Authority

Most owners delegate authority to the field personnel for approval of change orders up to a certain dollar amount. The specific amount is usually a function of the size of the project or the funding mechanism available to deal with payment of the project costs. When the owner's field personnel have no authority to deal with cost issues, and a contractor's requests for increases in its contract amount require approval of a Board or some other bureaucracy, which may meet very infrequently and may not understand the realities of construction, the project environment is ripe for disputes and claims.

2.5.4 Unreasonable Contract Interpretation

If the contract language is subject to more than one reasonable interpretation, a contractor's interpretation may often differ from that of the owner. If the owner unreasonably rejects the contractor's interpretation, or if the contractor's interpretation is unreasonable, a claim situation may develop to resolve these differences.

2.5.5 Inadequate Documentation of Performance

Owners and contractors must document the issues, costs, schedule details, and events that affect the project performance. Absent requirements or internal procedures to record and maintain this documentation, the facts that are necessary to resolve the claims and disputes that develop during the execution of the project are not available except in the heads of the personnel. An accurate recollection by project personnel of the facts pertaining to disputed issues during lengthy and complex projects is unlikely. Claims will develop to resolve whose version of the facts is the most persuasive.

2.6 Causes Associated with Claim Settlement Procedures and Practices

2.6.1 No Written Procedures and Process

When disputes develop that cannot be resolved at the project level, a mechanism should be identified in the contract documents to take the dispute to a higher level of resolution. Often, the contract documents identify forms of alternate dispute resolution, such as meetings of executive management of the parties, mediation, dispute review boards, or project neutrals, as a necessary step before binding processes such as arbitration or litigation. While these intermediate processes are often helpful and cost effective in terms of resolving the claims or disputes that develop, the contract documents should also refer to the agreed-upon procedures that the parties must follow while engaging these dispute resolution methods. Of particular importance in these procedures are the following:

- The format and contents of the contractor's claim, including the identification of schedule analysis methods and cost documentation;
- Whether discovery of information and documents will be permitted;
- Time frames for moving through the process; and
- Finality of the decisions.

2.6.2 Discouragement of Project-Level Settlements

Owners and contractors who encourage the field personnel to resolve the disputes at the project level and discourage the bumping up of problems to the head office are less likely to have claims and disputes requiring third-party resolution.

Chapter 3

Why Construction Disputes Occur and How to Promote Early and Amicable Settlements

Chapter 3

Why Construction Disputes Occur and How to Promote Early and Amicable Settlements

Most construction projects, irrespective of size, type, and location, are subject to claims and disputes of numerous kinds. Claims are not parochial by nature. They are not peculiar to specific legal systems, languages, or cultures, or exclusive to certain materials or processes. Construction is full of risks involving all stakeholders in a project. How and when these risks are identified and managed will impact overall project success and determine if each member of the team serves the mutual interest of all stakeholders and the project itself. Not all risks can be foreseen, and often it is the unforeseen impacts that create increased distress and conflict among the parties due to lack of understanding, insufficient planning, inadequate claims and disputes resolution processes, and inadequate management skills that result in an adversarial situation that can escalate exponentially without resolution, particularly where large sums of money are involved and neither party is willing to move from its stated position, notwithstanding the basis of merit.

The most effective way to deal with construction claims is through a common doctrine of prevention and/or early recognition and efficient change management utilizing solid contract documents that identify and allocate the construction risks between the parties in concise and defined language, understood and utilized by everyone. Unfortunately, such idealistic circumstances rarely, if ever, occur. However, even where such conditions do not exist, given the appropriate actions and application of contract administration processes by the parties involved, it should be possible to promote a set of circumstances that will help facilitate, if not early resolution, a general settlement later.

Construction claims can manifest themselves in numerous ways as combinations of key contributory factors as diversified as interpretation of contract documents, legal systems, cultural perspectives, statutory regulations, interparty communications, and even personalities. Disputes can arise at any time and any stage of the construction process from concept through bidding, engineering, procurement, construction and even post completion after the project has been handed over. Construction disputes are common, but much can be done to manage the consequences of such core issues as delay and disruption, particularly given that in today's world it has never been easier to "manage the risks of change and to prevent what, in the past, was thought to be the inevitable knock-on effect on completion."[8]

It is sometimes said that the collective noun for the parties to a construction contract should be called an "argument." It is unfortunate that such a comment all too often rings true. As soon as the parties have signed a contract, whether between the main contractor and owner, contractor and subcontractor, owner and supplier/vendor or design consultancy, or any other combination thereof, the complexity and often burdensome processes involved with individual party interpretation and application of contract language, drawings, specifications, statutory regulations, etc., begins and continues until the end of the project. It is this self-interested interplay between the various contractual relationships, thousands of supposedly coordinated construction documents, overreliance upon misunderstood or misinterpreted legal precedents and systems, and a combination of numerous other applicable factors, coupled with the commitment of the monetary risks and financial stakes involved, that provides a breeding ground for lack of trust and the rapid creation of an adversarial atmosphere between parties, often from the beginning of the project.

Notwithstanding the plethora of contributory factors that can and do create disputes on construction projects, there is, however, one factor that stands out regarding its ability to provide both a negative framework in the development of disputes and a positive foundation in helping resolve them, and that element is communication. Webster's dictionary defines "communication" in part as follows:

> *... the act or process of using words, sounds, signs, or behaviors to express or exchange information or to express your ideas, thoughts, feelings, etc., to someone else; a message that is given to someone: a letter, telephone call, etc.; communication: the ways of sending information to people by using technology.*

All too often, the lack of exchange of information and/or comments and concerns, or not communicating contemporaneously, or at all at any stage of the project, simply provides the basis for a dispute to crystallize and grow from what can seemingly be a benign and innocuous beginning into a complex, expensive and, on its face, irredeemable issue of contention with a life of its own.

Communication as one of the core drivers of and key reasons as to why construction claims develop is not, however, exclusive to the interface between the various contract parties involved in a construction project. It also relates directly to the individuals within a single party. For example, where the contractor's project team does not communicate well with its own upper management, this can lead to misunderstandings about, *inter alia:* 1) the real status of the project both in time and cost as opposed to the internally reported

---

[8] Keith Pickavance, *Delay and Disruption in Construction Contracts*, Third Edition, 2005, Introduction and Terminology, p. 1.

level, 2) the veracity and basis of merit of any changes and potential claims, and 3) commitment of management to support the onset of the adversarial process based on the wrong or poorly defined premise.

> *A primary difference between situations that avoid disputes on a project and those that do not is that disputes are avoided when the participating organizations and the project team working together have the ability to develop an open communication strategy regarding issues of concern to a participant and are committed to resolving issues in a proactive, cooperative, and timely manner as they arise.*[9]

Where good and detailed communication exists at all stages of a particular issue and throughout the project, and from all parties involved, it provides a solid foundation upon which to facilitate an early solution of a specific isolated problem or a documented basis for later resolution of much larger encompassing disputes. All parties involved are part of the communication process from the beginning. When good communication exists, the parties have a more realistic and better understanding of the risks and implications of the various contributory parts and facts of the dispute as presented, and it is clearly easier to reach an agreement based on timely and detailed information, even if it is questionable, than where the information is lacking or does not exist at all.

Timely and positive communication fosters the potential avoidance of disputes and claims, allowing the parties at least to attempt to resolve specific matters before they develop into more complex and larger adversarial issues. There is, of course, no guarantee even when excellent communication has taken place that resolution and settlement will occur without additional due diligence and assistance by third-party regulators. However, given the nature of the dispute process, whether it be negotiation, mediation, arbitration, litigation, or any other alternative dispute resolution (ADR) approach, the presence of a bulk of contemporaneous communication and materials can only help to bring parties closer together and provide more definition for a fair and reasoned judgment and award.

Numerous articles have been written that describe in great detail the plethora of causes of why claims occur. Accordingly, the intent herein is not to follow a list of contributory mechanical issues but to provide an outline of some of the lesser discussed reasons why claims and disputes occur during the construction process and to demonstrate that, by following a simple approach to the effective management of such issues, the opportunity to avoid claims, foster early resolution of changes and/or promote settlement of disputes can be greatly increased. The concepts involved are not complex and relate to common sense basic principles of project and construction management, which, unfortunately, are

---

[9] *Guide to Leading Practice for Dispute Avoidance and Resolution*, CRC Construction Innovation, November 2009, Managing Disputes, p. 28.

all too often found lacking even on major projects and among sophisticated global construction stakeholders.

In this section, we discuss issues causing claims during the following stages of a project:

- The RFP/bid stage
- Project execution
- Post-project completion
- The settlement process

3.1 The Request for Proposal (RFP)/Bid Stage

The process of preparing the construction contract, whatever the size, may be divided into three simply defined parts:

1. The decision of the type of contract and the terms and conditions under which the work will be carried out.
2. The selection of the contractor.
3. The establishment of the contract price or how the price will be determined.

3.1.1 Contract

If a standard contract form is not used, owners often develop their own ad hoc form. In this latter regard, owners can expressly attempt to redefine and reallocate all construction risks in their favor, often placing undue burden on the contractor. The ad hoc form can be poorly drafted and, therefore, create a myriad of problems related to lack of terms definition, misinterpretation, misunderstandings, and lack of legal guideline precedence. Furthermore, the over-apportionment or lack of definition of risk can cause a contractor to increase its price to a point of being noncompetitive, or failure by the contractor to recognize such risks at the bid stage can create disputes during the execution of the contract when the contractor attempts to recover its unanticipated risks as they transpire through the contractual change mechanisms.

It is essential, particularly where ad hoc contract forms are used, that experienced legal input is utilized to redline the contract for clarification, definition, and negotiation purposes. Redlining contract forms is a common practice that helps both parties to better understand the operational parameters of the terms and conditions. If a contractor does not clearly perceive the meaning of a term or allocation of a risk, it is again vital that this be raised and expressly clarified at the prebid and precontract stages; otherwise, it will almost certainly result in later conflict and disagreement.

The contract provides the legal basis of the obligations, rights, and entitlements of the parties, so it is crucial that both sides, where possible, achieve common ground on how the contract reads and is to be implemented. The only way to do this effectively is to raise questions through the prebid Q&A or redline process and seek at least clarification, if not joint agreement, regarding comprehension of specific clauses. Not to do so is a guaranteed recipe for a dispute to occur. Contractors either fail to do their due diligence with the prebid contract review or make unilateral assumptions about interpretations, both of which can prove to be costly.

One area where this approach cannot be applied is with Federal contracts under the Federal Acquisition Rules (FAR). The Federal government does not allow any changes to the contract form, and any participating contractor is expected to be entirely familiar with and understand the basis of the contract terms and conditions. This places a specific emphasis on the contractor to be thoroughly versed with all aspects of FAR, as no excuses to the contrary are acceptable at a government and/or supporting law level.

3.1.2 Contract Documents

If the RFP and/or contract documents (contract, drawings, specifications), are not properly prepared and coordinated, and/or if they contain errors, particularly of magnitude, this creates ample latitude for both parties to facilitate disputes during project execution. From the owner's side, the position is usually one of denial, placing the emphasis upon the contractor to resolve such document conflicts from an "assumed all risk" umbrella, particularly under a lump-sum fixed-price contract. The contractor adopts a contrary position of "defective specifications and drawings," which on its face has considerable case law and legal support, but this does not necessarily provide a timely solution to the discrepancy, which can result in an expansion of the dispute at a time when it needs to be resolved expeditiously for the benefit of the project. There is no shortcut regarding an owner expending time and money upfront to provide a properly coordinated and prepared set of RFP and contract documents. However, the rush to achieve a commencement deadline coupled with the quick and simple ability to electronically "cut and paste" from other projects can result in serious documentation errors.

Clearly, the requirement is upon achieving a fully coordinated and correct set of contract documents that benefits everyone involved in the project. However, there is no such thing as a perfect set of contract documents, as mistakes and errors of some form will always be present. The key is to minimize both the presence and effect of such issues through prudent and detailed reviews by the participating parties, as well as ensuring that the contract provides express procedures to be followed when errors are discovered and clarifies where the liability resides and who is responsible for the appropriate redeeming actions. Many ad hoc contract forms fail in this latter regard.

Contractors are sometimes afraid at the prebid stage to highlight document errors or discrepancies on the premise that it will somehow disqualify their tender or work against them in some way. However, such an approach places the contractor at risk of having assumed responsibility for the issue during construction if prior notice is not given during the bid stage.

It is not unusual for contractors to be placed under considerable pressure to review and issue their bid pursuant to a short tender period. If the contractor is not allowed sufficient time to properly review the RFP documents and facilitate a competitive price, two negative outcomes are likely to occur. One is that the contractor's price will be high to make sure that any risks that could not be fully ascertained during the bid process are covered by some level of built-in contingency. Secondly, the price may be too low for failure to recognize potential cost impacts associated with such risks. In both cases, the result is likely to end up in a dispute when the risk arises. If the contractor's price was high, the contractor may still attempt to claim for additional costs of the risk when it occurs irrespective of any internal contingency allowed for in its bid, unless the contingency itself is expressly identified in the bid. If the contractor's price was low, the contractor will fight to recover costs not allowed for in its contract price.

The assumption of risk is part of any construction contract. For the contract to be executed successfully, however, it is paramount that within the contract documents, all foreseeable risks are identified and allocated to specific parties, and management/action processes are identified. Similarly, for any unforeseen risk, the contract documents must also expressly identify the processes to be followed to manage such risks. If this is not done fully and correctly, the result will be a plethora of claims and disputes, with both parties fighting over contract language and risk interpretation. The contract must be expressly specific about the mechanisms to be followed by all parties when changes and risk events arise. Standard contract forms through committee development over many years provide solid frameworks; however, ad hoc contracts by their very nature as one-off forms may not. If the contract is lacking in this regard, the onus is upon the contractor to question such discrepancies during the prebid Q&A stage, requesting answers about interpretation of specific contract clauses, if not thoroughly understood, and/or lack of definition of risks and change management processes to facilitate potential conflict reduction.

A key purpose of drafting solid contract language is to avoid claims and disputes. Notwithstanding, whatever and however the contract defines the recognition and management of risk, it must provide mechanisms that allow the progress and continuing operation of the project, even if an issue is in dispute, to avoid costly delays if immediate resolution cannot be achieved.

3.1.3 Question and Answers (Q&A)

Most contractors fail to make sufficient use of the prebid Q&A opportunities prior to finalizing the tender price. Contractors however must never be afraid to use this stage of the process to raise any questions and concerns about the RFP, including but not limited to: 1) the interpretation, understanding, and operation of the contract itself, 2) risk and change mechanisms, 3) clarification of errors and discrepancies, and 4) any other issue that will help facilitate a competitive package. Furthermore, when a question is raised and a definitive answer is received, it provides an express framework for a specific issue that should help avoid the onset of a dispute if and/or when the matter arises during the project. At the very least, it provides a written record of the interface and potential agreement between the parties that is difficult to refute later.

The author has witnessed many instances when a sophisticated major contractor on a large, complex project, during the prebid stage, has failed to raise a single question regarding the RFP package or solicit any additional information that would help with its tender preparation and potentially prevent claims and disputes during construction execution. When a claim does occur, a key part of reviewing both liability and entitlement is to look back at the contractor's basis of bid and particularly the Q&A sequence prior to the bid and contract execution. If the contractor raises a question after contract award that could and should have been raised during the prebid Q&A stage, the potential for the presumption of risk being placed on the contractor is high.

Clearly, not all issues can be raised, questioned, or even recognized during the project bid stage; however, when more clarifications are sought by the contractor during this period, fewer issues of dispute will result later. A contractor may be disqualified if it attaches a list of prequalifications with its bid; however, no contractor can be disqualified for seeking clarification and asking questions during the prebid period.

3.1.4 The Bid

The lowest bid does not necessarily mean either a qualifying or solid basis of bid. If a contractor is desperate for work or simply decides to buy the contract for whatever reason, this is likely to facilitate a low bid that may not support the full execution of the project without recourse to claims for additional money and/or time. In the private sector, there are generally no specific rules or laws that require an owner to question the lowest bid except for several legal precedents regarding the correction of bid errors. In the government sector, however, under FAR, where there is a considerable price difference (usually 20 percent plus) between the lowest and second lowest bids, the government is under an express obligation to notify the lowest bidder of the disparity (without providing cost or quantum details) and request the lowest bidder to reaffirm or amend its bid. If the government does not do this during the bid and contract negotiation stages and awards the contract to the lowest bidder—which then discovers the disparity post award—there is legal precedence for the contractor to challenge the basis of its bid price and attempt to recover additional monies that it may or would have included if it had been advised of a bid level price disparity. Even if the contractor is not aware of the detail of the differences, it clearly understands that a 15 percent or 20 percent difference or more exists pursuant to the government's notice in this regard.

Where there are clear levels of pricing difference among bids, the owner should take appropriate steps to assess why such disparities exist and what action to take, if any, to verify the basis of the differences. For example, two or three bids that are close together may indicate a solid and reasonable contract price; however, a bid that does not conform to the group datum average, particularly if it is lower, may indicate a potential problem with the bid. Significantly low bids raise the question: Can the contractor execute the contract for the price quoted without recourse to a series of change requests and/or claims?

It is not always easy for owners to analyze or question such bid differences, particularly as raising a question about a basis of bid may well cause the low price to change upwards when the contractor is faced with either amending or reaffirming its bid. However, by simply asking the question of the contractor in this regard, the owner is potentially avoiding or reducing the onset of future claims facilitated through a low basis of bid or, at the very least, providing a much stronger foundation to rebut later associated change requests.

An owner is always drawn toward the lowest bid price for the obvious financial reasons; however, if it fails to undertake any bid relationship due diligence, it is likely to discover itself embroiled in a series of potentially avoidable disputes during project execution related to the contractor's bid price. The fact that a bid is questioned and/or reaffirmed during the pre-award stage is no guarantee that claims related to bid price will be avoided; however, it is always better to raise concerns and questions before and not have them surface after the contract is awarded. It is not an easy decision for an owner to question a low bid to the point of rejection. However, accepting such a bid can result in major disputes and the potential exposure to considerably more cost and expenditure in fighting and resolving the matter compared to an increased bid based on precontract discussions and negotiations.

In terms of a contractor buying a project, government Indefinite Delivery, Indefinite Quantity (IDIQ) contracts carry high potential for bid abuse. Under such an open operational framework, contractors will often low bid the first contract with the intent of getting a "foot in the door" and establishing a relationship with the owner or government entity to facilitate a chain of future contracts based on more realistic and profitable bid levels. However, the first low-bid contract, depending upon the bid/execution price disparity, may still result in the contractor fighting to recover lost monies, damaging contractor and client relationships for future work (the very opposite of what the contractor was attempting to achieve) and maybe even causing the contractor to overbid future contracts to recover the initial low bid losses.

3.1.5 Bid Qualification

Where the RFP stipulates that a contractor is not allowed to prequalify its bid, this can result in either the assumption of risk in the bid for specific issues and items that the contractor may not be prepared for and, hence, will fight to recover later, as and if such risks occur, or the bid price itself will be pushed upwards to cover such factors, sometimes to the extent of making the contractor less or even noncompetitive.

Government contracts usually do not allow contractors to prequalify bids. This can place the contractor from the onset of the award in a dispute mode questioning issues under the umbrella of "presumptive risk" based on the unqualified bid process and is certainly a major contributory factor as to why so many federal contracts end up in some kind of dispute forum. This is not so with the private sector. The very basis of a prequalification potentially clarifies areas of contract risk that need to be resolved and allocated prior to the contract award. Where both parties jointly review a specific item or series of prequalified risk factors prior to the contract award, the potential for a conflict arising from such matters during project execution is greatly reduced through express clarification and agreement.

Where prequalification of a bid is not allowed, and even under federal contracts, there is nothing to stop a contractor during the prebid Q&A stage seeking clarification of any issue or item accordingly. In fact, it is essential that the contractor follows this approach for the obvious reasons of contract clarity, basis of bid, and the potential avoidance of claims.

3.1.6 Project Budget

Where an owner has not completed its due diligence to ensure that the contract budget preparation matches the basis of the project design and execution, and when the budget is tight or no contingency is allocated, an increased potential for disputes and claims occurs. Thus, claims are a predicted outcome because the owner has little or no flexibility to approve required changes as the project develops.

It is essential that any construction project should have some level of contingency allocated to the budget process as projects are rarely, if ever, executed without some level of change occurring. If the owner cannot accommodate changes, the project will invariably suffer. Owners will want to facilitate changes for the benefit of the project because no set of contract documents are perfect regarding full coverage of the project requirements, and contractors will want to be paid for implementing such changes. Any kind of dichotomy between the two will result in claims and disputes.

The emphasis is upon the owner to solicit the appropriate funding before the project commences. While this may appear to be obvious, it is not always the case. Furthermore, if a project undertakes changes to the extent that the budget and/or contingency is used up, this creates a situation in which the owner has no ability to accommodate further changes or settle contractor change requests and claims, which will then create the onset of a dispute.

While owners have an obligation to facilitate appropriate funding to fully support project execution, a similar emphasis is placed on the contractor in terms of its bid preparation covering both foreseeable and sometimes unforeseeable risks, depending upon the risk allocation expressly described in the contract documents, through accurate cost analysis and the application of a contingency sum either internally allocated or externally and expressly declared.

Contractor-allocated contingencies are often the subject of disputes as to who benefits from their appropriation. For an externally declared sum identified within the contractor's bid, the owner will attempt to use the contingency for owner-initiated changes, thus accommodating the costs within the original bid price, while the contractor will want to expend such monies to cover the occurrence of risks that are within its contractual scope.

The process of managing contingency sums can create disputes between the parties. An internal and undeclared sum by the contractor as part of its bid is easier to manage but can result in problems regarding how much to allocate compared to the risks that may occur and in terms of facilitating a competitive bid.

An owner requirement for a contingency or provisional sum in the RFP and bid may or may not provide a better definition of its use and allocation, but in the absence of a clear and unequivocal contract stipulation in this regard, the process is open to potential dispute as to whose money it is.

Conflicts regarding use of contingency can be limited if all parties involved in the project agreement understand the purpose of the contingency and how it will relate to their respective roles during the project execution.

3.1.7 Communication

The RFP and bid stage of a construction project is full of potential pitfalls and risks that can result in a series of claims and disputes during the actual execution of the project. However, the above issues and processes have one clear and undisputed overriding common factor involved that helps to avoid and/or resolve the onset of disputes. That factor is consistent, express, and open communication between the parties at every stage of the project.

3.2 Project Execution

For the sake of brevity and clarity, project execution can be split into the three separate disciplines related to EPC work:

1. Engineering and design
2. Procurement
3. Construction

The above processes interface in different ways depending upon the type of contract utilized, *i.e.*, lump-sum turn-key as opposed to the more traditional design/bid/build. However, the fundamental issues found at the core of potential disputes resulting from the baseline disciplines are not dissimilar.

Most conflicts and disputes arise during the construction stage of a project. This is when previously unknown design, bid, and procurement issues usually manifest themselves as part of the execution process, when both time and cost factors assimilate themselves in a critical matrix that can cause the project to become embroiled in a series of claims and disputes, which if unresolved can severely and negatively impact all parties involved and the project itself.

There are a plethora of reasons and combinations of causes as to why claims and disputes occur during the project execution phase, and there are numerous articles that describe such factors in detail. The following sections highlight some of the key, fundamental, underlying, repetitive, and sometimes misunderstood foundations of why conflicts occur during the project implementation stage, based on the author's extensive experience on major global EPC projects.

3.2.1 EPC Format

Where an EPC contractor has sufficient contractual leeway to provide a design that satisfies the owner's RFP and/or performance specifications and its own basis of bid in terms of design and construction expertise coupled with a competitive tender price, and where some design assumptions may not have been declared during the bid process, there is always considerable risk that both parties have moved into a construction agreement whereby the expectations and requirements of the owner and contractor are at odds with each other.

Disputes can occur because of undeclared and misunderstood differences between the parties as the project moves from contract award into the design and detailed development stage. If the project is of a particular type, for example a power station or water treatment plant, where the contractor was selected based on its specific and perhaps extensive experience with similar projects, the contractor is likely to favor specific factors in the design and procurement process that allow it to be competitive, providing they are not in conflict with the basis of the RFP and any precontract negotiations and declarations. However, any assumptions and allowances the contractor has made in its bid do not always translate directly into owner expectations and needs.

A contractor will use standards, materials, and equipment that it expects will meet the RFP and the contract requirements. This allows the contractor to be competitive and provides a foundation for a planned, realistic, and understood framework for construction based on previous and similar experience. As the contractor submits the design through the various stages of the owner approval process, conflicts can occur and manifest themselves in two distinct ways:

1. As the design develops, both parties end up in a series of disputes that, if unresolved in a timely and positive manner, create project delays and foster mistrust and poor interparty relationship problems thereafter. This occurs when the contractor's submittals conflict with the owner's expectations, notwithstanding the basis of bid and any precontract interface. More often than not, the owner will not want to pay for any changes to the design as the owner deems them to be included in the tender price and believes it has the right to request or instruct changes on a "no cost" basis because "that was what they wanted in the first instance" even though the RFP may not have been specific or even declaratory in this regard.
2. Contractors accept minor owner requests and small changes during the design development and approval stages without fully understanding the potential time and/or cost ramifications that can result at a later time. This accords to the much-noted construction principle of "death by a thousand cuts." Contractors, particularly at the beginning of the project, will go out of their way to avoid upsetting the owner in this regard. It is all too easy for a contractor during the numerous approval and review stages to allow minor drawing and specification changes and developments, only to discover that it has utterly failed to comprehend the impact to the procurement and construction process. Depending upon the magnitude of the cost impact, the contractor will probably end up disputing the additional costs.

3.2.2 EPC Conflict of Requirements

If an owner fails to fully understand the basis of how a project has been bid under a turn-key EPC umbrella, for example, compared to its own desires and needs, claims and disputes will occur. If the RFP and contract documents do not fully reflect what the owner wants and/or expects, or give the contractor sufficient leeway to make certain allowances in its bid, the project is likely to suffer as any differences surface during the project development.

It is essential that open and continuous communication between the parties exists at all times. While not every contingent risk or change can be covered, the concept of avoiding future conflicts and disputes through an express interface process at each stage of the EPC process, particularly during the preparation and coordination of the design package, is fundamental to supporting the reduction of potential claims during construction.

3.2.3 EPC Coordination

Any significant lack of coordination between the parties at any stage of the project, even within each individual participating party's own organization, will result in disparities that foster change and create disputes.

In the rush to submit a successful bid, contractors can fail to pay the required attention to the contract start and completion dates specified in the RFP. If the contractor fails to analyze the realistic achievement of the contract period for the type, size, and format of the project to be built, no matter how well specified and detailed the design, the project is likely to fail because of unrealistic time constraints being imposed against the risk of liquidated damages for a late finish.

This is further exemplified if the contractor's detailed schedule lacks sufficient and careful coordination of items such as key procurement milestones against construction needs. The risks associated with key procurement items should be built into the main construction schedule as core activities, and this is particularly vital if the equipment or plant is being provided by the owner. While this may appear to be obvious, the author has witnessed many instances of such occurrences creating major project delays that could and should have been avoided with the application of a simple "common sense" approach and general due diligence by both parties.

A factor that is often missed by contractors is the risk involved with the procurement and delivery of a key construction item or piece of equipment compared to the overall risk of a resulting delay to overall project completion. For example, if a key item is procured from a supplier and is crucial to a project schedule, the contractor will correctly impose liquidated damages in the purchase order with the vendor to cover the contingency of delay. However, where the process may break down is the level of damages potentially levied against the vendor under the purchase order (PO) agreement compared to the contractor's exposure of LDs under the main contract. Where a major equipment vendor may well accept, for example, $3,000 per day in LDs for being late with its delivery, the contractor may be facing $25,000 per day under the main contract. It is unlikely that the vendor would accept LDs to match the main contractor's exposure. Thus, the contractor is placed at considerable risk in this regard. One way to attempt to avoid such an occurrence is for the contractor to manage the vendor supply on a critical day-by-day basis, perhaps having a contingency plan in place in the event a delay occurs. Another means of mitigation is to provide third-party insurance to cover any such contractual gaps. No matter what occurs, communication with both the vendor and the owner is vital to facilitate a possible workaround and alternatives. If the owner is part of the process through open communication, the opportunity to avoid LDs and end up in a dispute can be marginalized.

3.2.4 Compliant Notice

Failure to recognize an impact issue and provide timely notice pursuant to the contract can prove fatal regarding entitlement and the right to recover losses. Contractors must be aware of and apply the contract notice clauses. Otherwise, they run the risk of losing any entitlement to claim for additional cost and/or time, even if the basis of the claim has merit. This is probably the singularly most crucial but repetitive factor as to why contractors' claims can fail. Numerous articles have been written about the use and application of contractual notice compliance. The intent of mentioning this as an issue is simply to highlight the fact that contract notice expressly relates to the key and fundamental component of contractual communication between the associated parties.

3.2.5 Personalities

Effective communication between contracting parties and its potential for breakdown, abuse, and failure can be observed in a common trait and occurrence that is not covered by any express or implied clause of a construction contract. The concept is one of personality and the potential for clashes to occur that can negatively impact the project itself.

Because of the risks, high stakes, and overall investment of resources and finances involved in a construction project, the potential exists for individuals in a position of management power to quickly adopt an approach of individual and/or company preservation, ignoring the basis of the contract agreement and its fair and reasoned administration and application. Once this occurs, it is often not recognized until it is too late to avoid conflict, and even then, it is sometimes ignored when one or both of the parties have committed themselves to a particular position irrespective of what a reasonable interpretation of the contract may or may not be.

Such instances can occur at any stage of the EPC process; however, they are probably most prevalent on-site, where conditions of time and cost are at their most critical. Senior management oversight and site reporting provide the potential to control and resolve such matters before they spiral out of control. However, for this approach to work, it requires timely, regular, effective, and honest reporting and communication.

If such problems are caught early enough, the intervention of other uninvolved parties can provide an amicable resolution. Without such interface, a situation can quickly be created that fosters a non-cooperative atmosphere, negatively affecting everyone involved and impacting the success of the project. This can happen among personalities in multiple contracting parties as well as internally within any one party. Simple, constant, and open communication can only help and not hinder, a concept that is all too often found lacking.

3.3 Post-Project Completion

3.3.1 Balance of Power

Once the project is complete, the balance of power and leverage between the contracting parties changes, moving from the contractor to the owner. During construction, the contractor can apply some amount of leverage to facilitate and agree on changes. While the project is still ongoing, depending upon the circumstances involved, there is the possibility of achieving completion without a series of ongoing and protracted disputes. This can sometimes help promote flexibility toward the possibility of settlement. As soon as the owner has taken possession of the project, the ability to foster an agreement using the project itself as a basis of discussion is dramatically decreased, if not eradicated. The contractor has little left of substance to bargain with other than to be faced with the onset of expensive dispute resolution procedures to recover the monies to which it believes it is entitled.

Any claim or dispute that can be settled early, and particularly during construction, is likely to be a better all-around solution for both parties than allowing the matter to continue post-completion, even if one or both parties compromise regarding the nature and level of any such settlement.

3.3.2 Project Losses

When a contractor is faced with financial losses and the owner is faced with the possibility of paying additional and unbudgeted costs, the potential for the continuance of a dispute increases considerably. Neither party wants to concede its position, particularly when the monetary gap is substantial. This is when control of the disputed issues moves from site management to corporate, and the aspect of shareholder responsibility and profit and loss can override a rational settlement approach.

When potential losses are substantial, it is sometimes difficult for a party to differentiate its true exposure from one based on hopeful and wishful thinking. If company management is not fully aware of its true or reasoned exposure, it can often be misled by the project team in terms of level of expectancy and recovery and adopt an intransigent position based on incorrect information instead of reasonable negotiation. When company management has been involved in the project on a regular basis during execution, the prospect of such an occurrence is greatly reduced. Management often takes a cursory review of a project, perhaps one project of many, to satisfy whatever oversight procedures it deems are acceptable and relies upon the site team to provide and dictate the basis of a developing dispute. If the site team, for example, has made unreported mistakes that have contributed to the potential or actual losses or failed to grasp and better understand its specific entitlement and rights, such a situation does not usually foster a later and open admission of its errors to upper management.

Claims or unresolved contract changes should never be booked or allocated as company assets or revenue. Contractors desperate to support their bottom-line losses continue to implement such procedures to bolster the company books pursuant to an audit or satisfy bank loan oversight. This process is dangerous in several respects, not least of which is that if a settlement is achieved but the final figure is less than allowed for on the books, a loss is still recorded, and the contractor takes a financial hit. The very process of booking claims as revenue promotes unreasonable claim figures to improve the company financials as much as possible. This practice also reduces the contractor's ability to negotiate a reasoned settlement as anything other than a full recovery pursuant to the booked asset as a direct loss, even if the claim amount had little or no merit in the first instance.

3.3.3 Lawyers/Experts

Companies sometimes involve lawyers, often specialized in construction practices, too early in the resolution process. As soon as one side involves a law firm in the claim process, the opposing side is effectively forced to commit to the same. This then further commits the parties to increased expenditure and risk of recovery to continue with the dispute process, which in turn creates additional pressure both in terms of financial obligations and timing. While law firms provide a valuable service, the question is at what stage of the dispute parties should bring in legal counsel. If a law firm or expert is doing its job correctly, it should provide its client with an early assessment and analysis of success to enable a rational decision regarding dispute resolution. Notwithstanding, the introduction of legal teams can sometimes overcomplicate issues and create the proverbial drawing of "lines in the sand" and battle cries from both sides.

Failure to understand the settlement processes described expressly in the contract as well as the general dispute resolution procedures available to parties in conflict clearly works against both sides. This can be further exasperated by a general lack of appreciation and understanding of the nature of the claims involved, the true nature of entitlement and reasoned recovery, the "blinding" effect of financial losses irrespective of fault, and general failure to have a game plan toward settlement. All these factors involve the concept of project communication at all levels and timing to achieve the desired result of a reasoned and equitable solution that each party can understand, accept, and move forward with. Unfortunately, this concept is found lacking in many construction players, large and small, sophisticated, or otherwise.

3.4 The Settlement Process

The following is a list of some of the key concepts and actions between the parties to a construction contract that promote the increased potential for the settlement of disputes as and when they occur. The list and commentary are by no means exhaustive but provide a framework that, if implemented, will help build a reasoned pathway to project success and achievement of amicable completion.

3.4.1 Communication

Open communication is essential at all stages of a project. As described hereto, perhaps the primary reason for the onset of disputes and lack of early resolution of claims is the failure to communicate issues and problems until it is too late, and a dispute arises. The prospect of an open communicative forum is not always conducive to the construction process. However, even under the most extreme of circumstances, such an approach, no matter how disparate, can help provide potential grounds for resolution, if not at party-to-party level, by an independent third party. It can also increase acceptance regarding the expectancy of success or failure. Open communication promotes negotiation, which is first and foremost the primary approach to early settlement.

3.4.2 The Contract

Pursuant to the issue of communication, the contract is the primary document that must be followed to preserve the rights and entitlements of both parties to the agreement. If the contract mechanisms are not managed and operated throughout the project in a timely and efficient manner in accordance with the express terms and conditions, then this is a recipe for failure. Conversely, correctly following the contract increases preservation of rights and the opportunity to achieve success in a dispute. Contract compliance promotes conformity of "notice," which is a vital element to any change or claim process and further fosters the production of accurate and detailed documentation in support of a party's position. This in turn facilitates a factual and contemporaneous framework to a claim as opposed to a retrospective and less credible approach.

3.4.3 Contract Compliance

Communication on a constant and reasoned basis between the parties cannot be overstressed in terms of the ability to provide a foundation for negotiation and agreement, no matter what the basis of the dispute. There will always be disagreements and the seemingly intransigent taking of sides; however, the party that has followed the contract and provided fully supporting and timely documentation will always be in a better position to succeed in a dispute forum. Contract disputes in arbitration or litigation are based on the "balance of probability" and weight of evidence.

3.4.4 Dispute Boards

When possible, include for a dispute board or adjudication process in the contract agreement. This provides an ongoing and contemporaneous mechanism by which a dispute or claim that occurs during the execution of the contract can be quickly resolved before it gets out of hand. This allows both parties to move forward with the issue resolved, at least on a temporary and relatively immediate basis, to the benefit of the project in avoiding major delays during execution. If one party disagrees with the Board or Adjudicator's decision, it can always dispute it at the end of the project. However, such a decision by a qualified, jointly appointed expert during the heat of the project often provides clarification and crystallization to a party regarding the future chance of success of overturning the decision in another and later dispute forum. Such an approach is not without risk for either party but does help the ultimate goal of allowing the project to complete sooner rather than later in an improved atmosphere where the potential for mistrust, conflict, and the escalation of a dispute are, at least in the short term, put to one side.

3.4.5 Stepped Negotiations

Contracts sometimes call for senior executives from each side to get together in a stepped attempt to reach a solution when a dispute arises. This could prove effective if the personnel involved have not had a heavy or regular and active involvement in the project as this allows them to act more independently and openly during any discussions and negotiations. However, the timing and timeline of such an intervention where possible should be contemporaneous to avoid protracted discussions or one side simply dragging out the process through lack of interest in settlement for whatever reason, of which there are many. If possible, both parties in dispute should attempt to establish communication links at management levels to maintain a general interface and the continuance of baseline discussions.

3.4.6 Less Can Be More

If an offer of settlement is made, particularly during construction execution, the contractor should give careful consideration before accepting or rejecting the deal. Depending upon the circumstances, it is sometimes better to accept less money and/or time during the execution of the contract than to hold out and fight for more at a much later date. Cash flow can be both the contractor's nemesis and king. A carefully rationalized acceptance of an early settlement may be of extreme benefit compared to nothing at all and the prospect of further risk and expenditure going after more. Both parties will benefit from a fair and reasoned approach in this regard.

3.4.7 Realistic Expectations

Both parties must be realistic in their expectations of settlement success and exposure to risk. This, however, is easily said but harder to facilitate for many reasons, as described above. One method that can help a party in dispute to reach a reasoned and clarified position is to utilize an expert consultant and/or law firm to help review the basis of the dispute and produce an independent report. Often for a reasonable investment in terms of consulting and legal fees, the party is provided with a better understanding of its rights and entitlement in terms of cost and time. It may not want to hear what it is told, but its position will be further crystallized in terms of a decision about how and/or if to go forward with the dispute. An exhaustive analysis at the outset is not usually needed, and a simple, high-level overview of a claim by an expert coupled with an outline legal review, if needed, can provide enough foundation to plan the next move. Any such outside expert report can then form the basis of further and more detailed analysis and reports if needed. The party can then negotiate from a position with some level of confidence and avoid the pitfall of reliance upon its own staff, which often has a vested interest in maximizing any form of recovery, irrespective of merit.

3.4.8 Legal/Expert Advice

Legal and expert teams can provide further support to a party in dispute to help it better understand the resolution processes involved, whether for arbitration or litigation, and the general level of expectation for success or failure. This also helps in the preparation of a budget for the dispute, which must be planned on a realistic and rational basis against the value of the risk or claim.

3.4.9 Commitment to the Process

When a dispute arises and a decision is made to pursue claim preparation or defense, then the party in question must commit to the process. When disputes reach the level of full-blown arbitration or litigation, any kind of half-hearted approach and overall lack of commitment is destined to fail. Furthermore, no dispute resolution process, no matter how strong the evidence, no matter how positive the entitlement, and no matter the level of confidence in the merit, is guaranteed success. The process is flawed in this regard; however, the chance of success can certainly be improved by adopting many of the project concepts and rationales described hereto. A party in dispute must always be prepared to lose, suffer the resulting consequences, and live with them.

3.4.10 Dangerous Practice of Inflating Claims

A party must never inflate its claim exceeding any rationalized and provable level of cost and time with the intent of settling for less during negotiations and effectively attempting to recover its baseline figure. For example, a contractor inflates its claim by 50 percent and then hopes to settle for half the amount, thereby achieving its planned recovery goal. This almost never succeeds and is a dangerous concept. What the contractor will find out is that the inflated portion of the claim will often bring into question and damage the positive and entitled portion of the claim, reducing the prospect of a fair settlement or any settlement at all.

3.4.11 Mediation and More

Wherever possible during the dispute process, it makes sense for both parties as early as possible to attempt to negotiate. Where negotiations continue to fail, mediation is the next step. Mediation is relatively inexpensive compared to most alternatives and can be facilitated quickly if the parties agree to do so. What a successful mediation will do, even if both parties do not actually settle as a result of the mediation, is crystallize each side's position regarding entitlement and risk. A party may have believed it had a strong position only to be told by the mediator that this is not the case. Such a comment from a qualified construction mediator should rapidly bring a party back to a reasonable level of reality and expectation, which can only help toward facilitating a settlement. At the very least, it will help to bridge the gap between the parties that existed prior to mediation.

3.4.12 Post-Dispute Audit

Any party involved in a dispute of any kind, at any stage or level, should, post resolution, hold an internal analysis and review to try to make sure that the reasons why the problem arose in the first instance are not repeated for future projects. While this makes absolute sense, it is not something that most construction contract stakeholders apply or undertake. A post-completion project audit can only be positive for the future avoidance of claims. Contractors and owners should not, but do, make costly mistakes time after time, effectively starting from scratch in each case, failing to communicate to the next project team and general management. The provision of simple procedures based on the premise of effective and timely communications can help manage claims and disputes and improve the chances of a successful resolution.

3.5 Summary

Dispute resolution in construction contracts has significantly evolved in the last 20 plus years with the development and inclusion on a global basis of ADR clauses in standard forms of contract as well as ad hoc owner forms. There is considerable variance in the cost of each approach, with negotiation being the most cost-effective and reliable form, then in order of rising expenditure and risk is conciliation or mediation, followed by dispute board or adjudication, followed by arbitration, and finally the most expensive and time-consuming of all, litigation. Whatever the dispute mechanisms are as expressly defined in the contract, the parties should always be encouraged to adopt a staged approach at the onset of a dispute, *i.e.*, maximize efforts to negotiate, and where possible stepped negotiations from lower to upper management based on an agreed timeline. Once the dispute moves from a failed negotiation to the next level in the chain, costs begin to increase, financial commitment and risk increase, and the prospect of the parties "digging in" begins to assume a different and more ominous perspective.

The maxim "prevention is better than cure" will always be the case in construction contracts. Maintaining constant interparty communications as described hereto supports both ends of the spectrum in that it can promote early resolution through negotiation as well as provide detailed and documentary evidence through whatever cure process is utilized as a dispute progresses. The foundation of effective communication relates to "clarity of understanding" and involves both transmission and feedback, which can be verbal but, wherever possible, confirmed in writing to benefit the dispute process itself. If both sides successfully apply and manage interparty communications, the chances of facilitating an early and amicable settlement as and when a claim or dispute arises are greatly increased. Clearly, there are no guarantees in this regard, but the very nature of a regulated and well-implemented communication process coupled with the regimented contract forum provide a supportive framework for potential interparty agreement at every stage of the project, from conception through handover and post-completion interface.

There are no substitutes for the requirements of following the contract, providing correct and timely notice, and producing factual and timely documentation at each and every stage of the construction process. All of these are contractual actions, but they are also part of a critical communication framework that is an absolute requirement if a project is going to be successful. When a party in dispute conforms to this approach, it provides the party with a more informed, supportive, and confident foundation upon which to plan for resolution and commit to the dispute process, whatever it may be. Accordingly, with a more informed and better understanding of the issues and the potential for success or failure based on the above premises, a party can be flexible and constantly strategize and re-strategize at various levels as the process develops and fluctuates toward resolution. Whether the result of the dispute is success or failure, a party that has utilized a well-communicated and contract-compliant administration, sought appropriate advice from experts, and adopted realistic expectations of the risks based on the facts will be in an appropriate position to attempt reasoned settlement at each stage and accept the outcome without recourse.

Early settlement is never guaranteed. Indeed, any kind of settlement when a dispute arises is never guaranteed. What can be guaranteed, however, is that by applying simple contract administration procedures, always maintaining joint communication lines at all levels, and promoting constant interface between the contracting parties, the project overall will benefit, and the probability of being able to achieve settlement through negotiation without recourse to more expensive, at risk, and time-consuming ADR procedures is greatly improved. There are no "rocket science" principles involved in fostering construction dispute settlements. Often, all that is needed is a commonsense communicative approach that allows both parties to negotiate openly and realistically under an umbrella of fairness, reasonableness, and equity; in other words, adopting the basis and common intent of the contract itself.

Chapter 4

Assessment of Problems Associated with Poor Project Management Performance

Chapter 4

Assessment of Problems Associated with Poor Project Management Performance

4.1 Introduction

In this section, the following topics are discussed:

- Purpose
- What is a construction project?
- What is project management?

4.1.1 Purpose

Experts are often called upon to testify during mediations, arbitrations, and court cases on the impacts of deficient project management performance that have occurred during the execution of construction projects. These impacts may have been caused by either party to a contract, *i.e.*, the owner or the contractor, or, sometimes, a subcontractor or supplier.

This section is to provide an explanation of the methodology and resources that experts commonly use to derive their opinions on allocation of the responsibility for various problems, and the impact of those opinions on the parties' entitlement to damages or extensions of the time of performance allowed under a contract.

4.1.2 What Is a Construction Project?

For the purposes of this section, we will define a construction project by drawing on the Project Management Institute's (PMI's) definition of a project from *A Guide to the Project Management Body of Knowledge (PMBOK® Guide)*,[10] from PMI's further discussion of the uniqueness of construction projects provided in the *Construction Extension to the PMBOK® Guide*,[11] and from the author's experience.

---

[10] *A Guide to the Project Guide Management Body of Knowledge*, ANSI/PMI 99-001-2017, PMBOK® Guide, Sixth Edition, 2017, Section 1.2.1, Projects, p. 4.

[11] *Construction Extension to the PMBOK® Guide*–2016 Edition.

PMI defines a project as a "temporary endeavor undertaken to create a unique project or service." PMI then provides examples such as "developing a new pharmaceutical compound for market," "improving a business process within an organization," "constructing a building," and others.[12]

In the *Construction Extension to the PMBOK® Guide,* PMI expresses the uniqueness of construction projects as follows:

- Construction projects should simultaneously address the geography, site conditions, communities, physical environments, existing infrastructure, as well as a wide range of stakeholder requirements.
- Construction projects often result in a one-of-a-kind product rather than mass-produced products.
- There is no opportunity to produce a prototype.
- A construction project may sometimes be performed in phases in order to provide an opportunity to review and refine the project design and implementation strategy, as well as validate the investment intention.
- Construction projects occur in an ever-changing, complex environment, and often with a high degree of risk.
- Construction projects often require the integration of engineering disciplines (civil, structural, electrical, mechanical, geotechnical, etc.) as well as interaction with technology and sophisticated equipment that demand unique construction techniques and methods.
- Construction projects typically carry large penalties or damages for projects that are completed late.[13]

Construction projects are also prime examples of projects that require progressive elaboration. According to PMI:

> *Progressive elaboration is a characteristic of projects that integrates the concepts of temporary and unique. Progressive elaboration means developing in steps, and continuing in increments.*[14] *For example, the project scope will be broadly described early in the project, and made more*

---

[12] We recognize that there are other definitions of a project, such as that shown in Cleland and King's *Project Management Handbook,* which states that a project is "*a group of tasks/activities, to be performed in a definable period, to meet a specific set of objectives.*" Government agencies, union agreements, and others also provide definitions of what constitutes a construction project when it is necessary for their purposes.

[13] *Construction Extension to the PMBOK® Guide*–2016 Edition, Section 1.

[14] From the *American Heritage Dictionary of the English Language,* 3rd ed. 1992.

> *explicit and detailed as the project team develops a better and more complete understanding of the objectives and deliverables....*
>
> *Progressive elaboration of a project's specifications needs to be carefully coordinated with proper project scope definition, particularly if the project is performed under Contract. When properly defined, the scope of the project—the work to be done—should be controlled as the project and product characteristics are progressively elaborated....*[15]

PMI further provides an example of progressive elaboration by referring to the development of a chemical process plant:

> *Development of a chemical process plant begins with process engineering to define the characteristics of the process. These characteristics are used to design the major processing units. This information becomes the basis for engineering design, which defines both the detail plant layout and the mechanical characteristics of the process units and ancillary facilities. All of these result in design drawings that are elaborated to produce fabrication and construction drawings. During construction, interpretations and adaptations are made as needed and subject to proper approval. This further elaboration of the characteristics is captured by as-built drawings, and final operating adjustments are made during testing and turnover.*[16]

The discussion about progressive elaboration is important, especially for construction projects, because the elaboration steps are carried out over an extended period of time, often for years on large projects. One of the challenges that project management faces in this regard is to maintain the consistency of the project's scope throughout project execution. This challenge continues through all the phases of the project life cycle.

The life cycle phases of a construction project, using an EPC contract as an example, typically include:

- The Conceptual Phase, including the owner's development of the project's first goals, design concepts, initial design calculations, and order of magnitude cost estimates, identification of financing mechanisms, taking the pulse of the public and regulatory agencies on the concept, etc.
- The Preliminary Phase (sometimes called the Basic Design Phase), including preliminary design calculations, preliminary plot plans, basic design drawings, development of permit packages, filing major permit applications, development of a contracting strategy, obtaining approval of major financing, ordering of long-lead-time equipment, and development of an EPC contract and bid package.

---

[15] *PMBOK® Guide-Third Edition*, 2004, Section 1.2.1.3, Progressive Elaboration.

[16] *Ibid.*

- The Bid Phase (or Contracting Phase), including identification of qualified bidders, solicitation and analysis of bids, clarification of the bids, and negotiation and award of the EPC contract.
- The EPC (or Execution) Phase, including all detailed engineering, procurement and construction activities, project management activities identified as the contractor's responsibility, pre-commissioning and startup activities, and possibly commissioning and initial operations activities.
- The Commissioning and Initial Operations Phase, if not included above.
- The Warranty Period.

Construction projects may develop grass-roots facilities, or they may remodel, refurbish, or expand existing facilities. Routine maintenance operations may also be combined with construction projects.

4.1.3 What Is Project Management?

Again, PMI provides a good starting point for a definition of project management:

> *Project management is the application of knowledge, skills, tools, and techniques to project activities to meet project requirements. Project management is accomplished through the application and integration of the project management processes identified for the project. Project management enables organizations to execute projects effectively and efficiently.*[17]

> *The project manager is the person assigned by the performing organization to lead the team that is responsible for achieving the project objectives.*[18]

It is important to note that many of the processes within project management are iterative in nature. This is in part due to the existence of and the necessity for progressive elaboration in a project throughout the project life cycle. That is, as a project management team learns more about a project, it can manage to a greater level of detail.

The author also recognizes that there are other definitions, some of which refer to project management as a "set of principles," but other definitions generally use somewhat different phraseology to say the same thing.

---

[17] *PMBOK® Guide–Sixth Edition*, 2017, Section 1.2.2. p. 10.
[18] *Ibid*, Section 3.2, p. 52.

4.2 The Importance of Sound Project Management Practices

The application of sound project management practices provides construction project stakeholders with the means to meet their objectives. These include ensuring that the project, *i.e.*, the building or facility is:

- Constructed to meet its intended purpose, and only its intended purpose;
- Constructed to meet the level of quality that was intended;
- Completed when it is supposed to be done;
- Completed for its intended cost; and
- Completed safely and while protecting the environment.

The lack of sound project management by owners or contractors on projects leads to construction delays and extra costs for both parties. In addition to the problems that occur during construction, poor project management can also result in a completed facility that fails to meet the specified quality and suitability of materials, fails to produce the intended products, or cannot be operated for its intended life. This usually results in claims from contractors and counterclaims from owners, or vice versa depending on who decides to sue whom.

Reasons for project failure that are often cited during disputes include:

- The failure of the project management team to adequately plan the work, or, when a plan is developed, to properly execute that plan;
- The failure to provide adequate human resources, staff, or direct labor, to the project;
- The failure to develop adequate project schedules, or to maintain those schedules throughout project execution; and
- The failure to control costs and changes throughout the execution of the project.

The above failures can be attributed at some level to inadequately trained or under-resourced project management teams.

In today's business environment, the application of sound project management principles on a project is not only the means to the ends listed above, but, during disputes, can also provide a level of defense against alleged management deficiencies that resulted in delays and additional costs. Demonstrating that you used sound project management practices, and that the other party did not, can help your side win disputes.

4.3 Project Management Requirements Stated in the Contract

Project management requirements are usually prescribed in major engineering and construction contracts, particularly in large EPC and lump-sum turn-key (LSTK) contracts. These contracts typically state each party's project management duties. These duties include requirements for development, review, approval, and implementation of:

- General project execution plans and additional detailed supporting plans for each management area
- A statement of the scope of work for the project
- Scheduling requirements naming the types and frequencies that various schedules must be produced and how they are to be updated
- Change Control
- Cost control and reporting
- Quality management
- Staffing and hourly labor manpower levels
- Performance reporting
- Insurance and bonding
- Subcontract management
- Management of materials and permanent equipment
- Job site safety
- Protection of the environment
- Claims procedures

During a dispute, a contractor's failure to perform its duties under the above requirements may result in loss of entitlement to any time extension or damages attributable to those failures. Similarly, an owner's failure to perform its duties may result in its inability to defend against delay and additional cost claims attributable to those areas.

4.3.1 Examples of Project Management Requirements from an EPC Contract

As an example of the typical duties stated above, an EPC contract executed between an owner company and a contractor included the following requirements under "Contractor Responsibilities and Representations:"

> *3.2.3 Required Skills and Capacity to Perform*
>
> *It (the Contractor) has the required skills and capacity to perform, and shall perform, the Work in a professional manner utilizing sound engineering, procurement and construction principles, current management information systems and electronic communication systems, project management and supervisory procedures in accordance with the Agreement, Engineering Record, and accepted practices in the industry. Its key personnel are as identified in Exhibit 8. (Emphasis added.)*

Note that even in a large EPC contract that one might consider to be all-inclusive in its internal statements of scope and technical specifications, the requirement to use "accepted practices in the industry" still exists. This statement, unless further defined to the satisfaction of the parties, can result in misunderstandings that lead to disputes.

Another EPC contract involving the addition of power cogeneration to an existing plant identified the following project management-related documents that the contractor had to produce during the work. These may be called "deliverables" or "submittals."

- Plans and Procedures
  - Project Execution Plan
  - Staffing Plan
  - Change Control Procedures
  - Communications Procedures
  - Procurement Plan
  - Quality Assurance Program
  - Transportation Plan
  - Construction Plan
  - Tie-in Plan
  - Shutdown Plan
  - DCS Training Plans
  - Performance Testing Plan
  - Turnover Plan
  - Startup and Pre-commissioning Plan
  - Surplus Materials Plan
- Safety Procedures Manual
- Schedules
  - Project Master Schedule
  - Construction Schedule
  - Steam Tie-in Schedule

  - Power Tie-in Schedule
  - Equipment Movement Schedule
  - Mechanical Acceptance Schedule

- Payment Schedules
- Insurance Certificates
- Progress Reports
- Proposed Products List
- Samples
- Manufacturer's Instructions
- Manufacturer's Field Reports

The contract mentioned above also noted a specific and rather extensive definition of the owner's project management role in the review and approval of these submittals, as well as for other contract requirements. The owner was required to review or approve documentation in the 23 areas shown below:

1. Quality Assurance Programs
2. Removal of Key Personnel
3. Disposition of Surplus Items
4. Site Cleanup Program
5. Subcontractors and Suppliers Used in Connection with the Work.
6. Gas Turbine Generator Availability Study
7. Master Schedule
8. Acceleration Plans
9. Mechanical Acceptance Certificate
10. Commissioning Methods and Procedures
11. Commercial Acceptance Certificate
12. Final Acceptance Certificate
13. Payment Milestones
14. Material and Equipment Requisitions
15. Invoice Payment
16. Subcontractor and vendor warranties and guarantees
17. Changes in the Work
18. Contractor-Furnished Insurance Certificates
19. Assignments (*e.g.*, subcontracting part of the Work)
20. Contractor Negotiations with Governmental Agencies

21. The Use of Photographs of any Company Facility
22. Public Relations Communications
23. Accounting Records and Right to Audit

One of the contractor's key arguments in a dispute related to the above contract was that the owner had overstepped the bounds of what the contractor argued was the owner's "limited" project management role by interfering with the contractor's means and methods of execution. This argument can often be defeated by demonstrating that the owner performed in accordance with the clear language of the contract.

4.3.2 Clarity and Support of Contract Requirements

It is imperative that both parties understand contract requirements. Because major construction project owners are typically the contract authors, contractors often attack the validity of contract requirements based on alleged unclear language and a resulting different interpretation of the requirements than the owner expresses. It is an advantage to the owner, therefore, to use clear language that can also be supported by generally accepted standards of project management performance.

PMI, the Construction Industry Institute (CII), and other respected institutions have created a large body of work that contract authors can reference to support the project management requirements shown in them.

4.4 Project Management Practice Generally Accepted in the Industry

No contract can address every eventuality that will occur during the execution of a project. When disagreements occur, one party or the other will often claim that what it was required to do or how the other party behaved was not in accordance with "normal industry practice" or "accepted practices in the industry" as discussed above. When this occurs, there are resources available to aid in the determination of what "normal industry practice" is.

When evaluating a party's project management performance that the contract does not explicitly define, experts can compare that performance with practices published by well-known and respected institutions such as PMI, CII, and others, as well as their experience.

4.4.1 PMI's *PMBOK® Guide* and the *Construction Extension*

PMBOK® divides the project management functions into ten knowledge areas,[19] and the *Construction Extension* adds two more areas particular to the construction industry.[20] Combined, these are listed in Table 4-1 below:

Table 4-1: Ten Project Management and Two Construction Extension Knowledge Areas

- Project Integration Management
- Project Scope Management
- Project Schedule Management
- Project Cost Management
- Project Quality Management
- Project Communications Management
- Project Resource Management
- Project Risk Management
- Project Procurement Management
- Project Stakeholder Management

CONSTRUCTION EXTENSION

- Project Health, Safety, Security, and Environ. (HSSE) Management
- Project Financial Management

Project Integration management, for instance, is made up of seven processes, including:

- Develop Project Charter
- Develop Project Management Plan
- Direct and Manage Project Work
- Manage Project Knowledge
- Monitor and Control Project Work
- Perform Integrated Change Control
- Close Project or Phase

While application of all project management processes associated with these knowledge areas is needed for the success of a project, some may not be explicitly specified in a given contract. Financial Management, for instance, may be left for the individual parties to handle apart from the contract. The owner would have to acquire project financing, make sure that funds could be distributed as needed, and handle accounting for capitalization purposes, but the contractor would not be involved in those processes. Conversely, the contractor would have to acquire its own financing for its needs and control those funds. The fact that requirements for financial management are not addressed in the contract does not mean that Financial Management is not needed.

[19] *PMBOK® Guide, 6th Edition,* 2017, Section 1.2.4.6, Project Management Knowledge Areas, pp. 23–24.

[20] *Construction Extension to the PMBOK® Guide*–2016 Edition, Sections 14 and 15.

Quite the opposite, a failure in Financial Management can mean the disastrous failure of the project.

Some contracts, lump-sum in particular, tend to lack detail in the areas of risk management, communications management, procurement management, and integrated change control processes. This leaves a lot of room for disagreements in these areas; therefore, an outside standard must be relied upon to determine which party's point of view is right.

The *PMBOK® Guide* and the *Construction Extension* provide an excellent structure for general project management requirements. During dispute resolution, an expert must be able to attribute responsibility for project management failures and to determine who is entitled to recovery of additional costs and time because of a party's failure to implement the project management knowledge areas and processes.

When the parties have implemented the basic project management processes discussed above and there are still disagreements, a more detailed analysis of the processes and how they were executed is needed.

4.4.2 CII's Knowledge Structure

One part of this more detailed analysis is comparison with the best practices available and recognized by the construction industry. One source of best practices is CII, an organization of approximately 90 companies and government agencies that are owners or contractors on the largest and most complex construction projects executed in the U.S. and internationally. Their scope covers oil and chemical companies, aerospace, food, timber, pharmaceuticals, power, and others. CII sponsors the basic research for and the development of "Best Practices" and other helpful supporting information and has organized this work into a "Knowledge Structure."

CII has organized its Knowledge Structure as a matrix that presents detailed CII publications by topic. These are individual instructive publications, *i.e.*, "how-to" books, called "Best Practices," "Proposed Best Practices," and "Information." The Knowledge Structure is divided into thirteen major "Knowledge Areas," all of which are focused on construction projects. The CII Knowledge Structure and publications are available when needed to both expand the scope of our project management analyses and to lend more detailed support to our findings.

CII's 13 Knowledge Areas are listed in Table 4-2:

Table 4-2: CII's 13 Knowledge Areas

1	Front-End Planning	8	Project Processes
2	Design	9	Project Controls
3	Procurement	10	Contracts
4	Construction	11	Safety, Health and Environment
5	Startup and Operation	12	Information Management and Technology Systems
6	People	13	Globalization Issues
7	Organization		

CII's Knowledge Structure showing the individual Best Practices, Proposed Best Practices, and Information publications is shown in Table 4-3 below.

Experts are often able to use CII best practices as support for their expert opinions related to construction scheduling, change order management, assessment of the impact of change orders on the cumulative impact of delay, and assessment of engineering and construction labor productivity. Publications that can be relied upon include:

- Publication 6-10: "The Impacts of Changes on Construction Cost and Schedule"
- EM113-22: "Scope Control and Change Management"
- Research Summary 158-1: "Quantifying the Cumulative Impact of Change Orders for Electrical and Mechanical Contractors"
- Source Document 108: "Quantitative Impacts of Project Change"
- Special Publication 43-1: "Project Change Management"

Table 4-3: CII's Knowledge Structure

CII's Knowledge Structure			
Knowledge Area	Best Practices	Proposed Best Practices	Information
1. **Front-End Planning**	1.1 Pre-Project Planning 1.2 Alignment	1.3 Early Estimating 1.4 Modularization/ Preassembly *See also* 13.3 International Project Risk Assessment	—
2. **Design**	2.1 Constructability 2.2 Design Effectiveness	*See* 5.2 Design for Maintainability 8.6 Work Process Simulation 11.3 Design for Safety	2.3 Piping Design 2.4 Design Standards 2.5 Cost Effective Engineering *See also* 12.3 Computer-Aided Design/ Drafting Process Industry Practices (PIP)
3. **Procurement**	3.1 Materials Management	—	3.2 Supplier Relationships *See also* 12.1 Automated Identification

CII's Knowledge Structure (continued)			
Knowledge Area	Best Practices	Proposed Best Practices	Information
4. Construction	*See* 1.2 Alignment 2.1 Constructability 3.1 Materials Management 5.1 Planning for Startup 7.1 Team Building 7.2 Partnering 8.1 Quality Management 8.2 Implementation of Products 8.5 Benchmarking and Metrics 9.1 Change Management 10.1 Disputes Prevention and Resolution 11.1 Zero Accidents Techniques	*See* 6.1 Employee Incentives 7.4 Leader Selection 8.4 Small Projects Execution 8.6 Work Process Simulation 10.2 Project Delivery and Contract Strategies 11.4 Managing Workers' Compensation	4.1 Competition *See also* 3.2 Supplier Relationships 6.2 Management of Education and Training 6.3 Attract and Maintain Skilled Workers 6.4 Craft Productivity 6.5 Multiskilling 8.3 Lessons Learned 9.2 Work Packaging 9.3 Cost & Schedule Control 10.3 Use of Project Incentives 10.4 Risk Management 12.1 Automated Identification 12.2 Electronic Commerce 12.3 Computer-Aided Design/ Drafting 12.4 Fully Integrated and Automated Project Processes (FIAPP) 12.5 Wireless Technology 12.6 Automation and Robotics 13.1 International Standards 13.2 Global Construction Industry
5. Startup and Operation	5.1 Planning for Startup	5.2 Design for Maintainability	—

CII's Knowledge Structure (continued)			
Knowledge Area	Best Practices	Proposed Best Practices	Information
6. People	—	6.1 Employee Incentives	6.2 Management of Education and Training 6.3 Attract and Maintain Skilled Workers 6.4 Craft Productivity 6.5 Multiskilling 6.6 Engineering Productivity Measurement
7. Organization	7.1 Team Building 7.2 Partnering	7.3 Organizational Work Structure 7.4 Leader Selection	7.5 Project Teams
8. Project Processes	8.1 Quality Management 8.2 Implementation of Products 8.5 Benchmarking and Metrics	8.4 Small Projects Execution 8.6 Work Process Simulation	8.3 Lessons Learned 8.7 Technology Implementation 8.8 Value Management
9. Project Controls	9.1 Change Management	—	9.2 Work Packaging 9.3 Cost & Schedule Control
10. Contracts	10.1 Disputes Prevention & Resolution	10.2 Project Delivery and Contract Strategies *See also* 13.3 International Project Risk Assessment	10.3 Use of Project Incentives 10.4 Risk Management
11. Safety, Health, and Environment	11.1 Zero Accidents Techniques	11.3 Design for Safety	11.4 Managing Workers' Compensation

CII's Knowledge Structure (continued)			
Knowledge Area	Best Practices	Proposed Best Practices	Information
12. Information Management and Technology Systems	—	—	12.1 Automated Identification 12.2 Electronic Commerce 12.3 Computer-Aided Design/ Drafting 12.4 Fully Integrated and Automated Project Processes (FIAPP) 12.5 Wireless Technology 12.6 Automation and Robotics
13. Globalization Issues	—	13.3 International Project Risk Assessment	13.1 International Standards 13.2 Global Construction Industry

4.4.3 Other Resources

Beyond those discussed above, there are other organizations that publish "best practices" and other resources that can be used to compare the actual project management practices an owner or contractor may have used on a project. Other organizations, for example, include:

- AACE International, which provides "Professional Practice Guides" such as "Contract Change Management – As Applied in Engineering, Procurement, and Construction,"[21] other publications on cost management, and a certification program for "Cost Consultants," "Cost Engineers," and other specialties.
- American Institute of Architects (AIA), which provides contract forms used widely throughout the building construction industry as well as publications on the management of design-build services and others.

[21] "Contract Change Management – As Applied in Engineering, Procurement, and Construction," AACE International Recommended Practice No. 100R-19, February 27, 2020.

- Center for Business Practices (CBP), which sponsors benchmarking forums and publishes resource books.[22]
- Construction Management Association of America (CMAA), which offers various construction management-related publications and a construction manager certification program.[23]
- Project Management Institute (PMI), which has a Scheduling Excellence Initiative (SEI), a two-year program organized by PMI's College of Scheduling. SEI is attempting the first global standardization and improvement of scheduling services and products since the inception of CPM nearly 50 years ago.

4.5 Measuring Conformance to the Project Management Requirements of the Contract

When construction management problems are part of a dispute, one needs to determine which party is responsible for each of those problems. As discussed above, project management requirements are normally laid out in the contract documents. The first step is to identify those requirements and create a compendium of them against which to compare alleged failures.

4.5.1 Deliverables

Using a contractor's duty to provide certain deliverables, for example, we would create a document that identifies the detailed requirements for each deliverable. The requirements may identify both the necessary content of the deliverable and when it needs to be provided. Certain examples of deliverables are identified in the sections below.

4.5.1.1 Project Execution Plan

The contract may require that a Project Execution Plan be provided along with the contractor's bid and then be updated periodically throughout the contract period. This plan typically contains statements of the project's goals and objectives; the project management approach; scope; budgets and project control methods; performance

[22] The CPB describes itself as "a knowledge center that captures, organizes, and transfers business practice knowledge to project stakeholders to help them excel in today's rapidly changing business world."

[23] CMAA also states that "CM is often used interchangeably with Project Manager." The CMAA is open to member "firms and individuals sharing a common interest in the promotion of construction management as a professional service."

measurement; preliminary schedule and major milestones; staffing and other resources needed; risk management; and supporting plans that need to be developed.

Once the requirements are identified, the actual plan that the contractor produced can be easily compared with the contract requirements to determine if they were met. This comparison requires not only that the original plan be reviewed, but also that subsequent updates were created according to the timing shown in the contract, and when significant events demand that the plan be changed to meet new or changed requirements. A poorly done or untimely plan is an indication of deficient project management.

4.5.1.2 Project Schedules

EPC contracts typically require project schedules be developed using the CPM method, so particular computer software such as Primavera Project Planner (P3 or P6) or MS Project is often used. An excerpt from the schedule requirements of an actual contract states:

> *4.6 Master Schedule. Within 30 days following the Contractor's receipt of the Notice to Proceed and in accordance with the provisions of this Agreement Contractor shall submit to Company, for Approval, a detailed schedule showing all Work Contractor proposes to perform during the first 90 days following said notice; on or before the end of said 90 day period Contractor shall submit to Company, for approval, a detailed schedule (including critical path networks covering each task of the Work including delivery of all equipment and materials), showing the tasks and related sequence of tasks Contractor proposes to perform the Work and achieve the requirements of this Agreement, including the schedule requirement of this article. Following Approval by Company this schedule shall be the Master Schedule. Thereafter, Contractor may present to the Company proposed amendments to the Master Schedule in a form satisfactory to Company and shall furnish in support of such proposed amendments such additional details and information relating to the Master Schedule or the Work as Company may request. Any proposed amendment shall be effective only upon Approval. Following Approval the amendment shall be included as an appendix to Exhibit 3. The Contractor shall implement the Master Schedule to minimize or eliminate the impact of any delay and to facilitate the orderly, efficient progression of the Work.*
>
> *In addition to the requirements shown above, the specifications of this Contract also required that the schedule be developed in accordance with a particular work breakdown structure, that schedules for mechanical acceptance and commissioning be developed at particular times, that updates be presented to the owner monthly, and that the updates be presented showing multiple levels of detail (e.g., Level 1–Summary, Level 4–Detailed).*

Like the project execution plan, the project schedule can be evaluated to determine its compliance with contract requirements. An analysis that shows that the schedule and its updates were not done when required, did not include all the elements of the work, and had defective logic or other errors would show that this contractor failed to properly schedule the project. This would demonstrate a significant failure in time management.

The contract quoted above also contained a provision that required the contractor to adhere to the schedule and, if problems occurred, to add manpower or take other measures to get back on schedule. Even if the contractor's scheduler had developed the schedule correctly, the field staff's failure to use it is a strong indication of deficient project management.

4.5.1.3 Progress Reports

The specifications of the contract discussed above required that the contractor provide a monthly narrative progress report that presented problems areas, anticipated delays and their anticipated impacts on the schedule, recommended corrective actions and their intended impacts, and the effects of changes on the schedules of other subcontractors.

Again, the progress reports can be reviewed to determine if they contain the type of information required, if they were published when required, and if they were accurate. The accuracy of progress reports can be determined by comparing them with project correspondence, meeting minutes, memoranda, and supervisors' logs that were prepared contemporaneously with the work. Any disconnect between these documents is an indication of poor communication management on the job, or sometimes, unwillingness to admit the truth on the part of those writing the reports.

A failure to properly develop the above deliverables can be used to show that a contractor failed to properly plan, schedule, or control the project from the very beginning. This failure can lead to the loss of his entitlement to a schedule extension or damages.

4.5.2 Contemporaneous Documents

In addition to the above, contemporaneous project documentation can be analyzed to determine whether the parties executed the project in accordance with the contract requirements. This documentation includes:

- Project plan updates
- Schedule updates
- Periodic reports
- Correspondence
- Requests for information
- Change orders (approved or not)
- Superintendents' logs

- Email
- Other contemporaneous documents such as photographs, telephone call records, etc.

The following examples indicate how these documents may show project management deficiencies:

- Where project plans have shown the staff and hourly labor requirements for the project, periodic reports and correspondence have shown that the contractor did not provide sufficient manpower to meet the planned levels.
- Progress reports have shown that the contractor did not place orders for material in accordance with the schedule, with the schedule in turn showing that this failure impacted the critical path.
- Change orders have not been developed in accordance with the detail and timeliness required by the contract and have indicated that the contractor agreed that there would be no effect on the contractual completion date as a result of performing the changed work.
- Superintendents' logs have identified the contractor's own deficiencies that led to delays or additional costs;
- Correspondence has shown misunderstandings among subcontractors that led to conflicts on the job site that negatively impacted worker productivity.
- Requests for Information have shown that the owner did not properly respond in a timely manner to the contractor's request for clarification of potential piping and structural steel interferences in the design drawings, leading to delay in the performance of the contractor's work until the design problems were corrected.

4.6 Measuring Conformance to Generally Accepted Project Management Practices

If an issue is raised that is outside of the specific language of the contract, the same documents listed above (project plans, schedules, etc.) can be analyzed to compare the parties' performance to the generally accepted project management practices discussed above. For example, material delays and change requests are discussed below.

4.6.1 Material Delays

If the contract did not require a specific type of procurement plan or status report from which to evaluate the contractor's performance, did the contractor generate its own procurement plan, schedule, etc., in accordance with a standard such as PMI's project procurement management standard? Provisions of this standard include development of a plan that shows the scope of the procurement effort, the required resources (*e.g.*, people), market conditions, constraints on availability of funds, and market assumptions. Procurement activities also must be shown on the project schedule to be properly integrated with the rest of the work.

A failure on the contractor's part to properly plan its procurement work, even if the contract did not require a plan as a deliverable, would be an indication of deficient procurement management. Any resulting delays and extra costs would be the contractor's responsibility.

4.6.2 Change Requests and Change Orders

If the contract did not specify a particular format for development of change orders, did the contractor provide an adequate change tracking system, identify personnel authorized to approve the contractor's change requests, identify the impact of the change on the project schedule, etc., in accordance with PMI's Integrated Change Control standard?[24]

A failure on the contractor's part to develop and control change orders in accordance with the steps noted above would be an indication of deficient change order management. Typically, this failure results in disputes between the owner and contractor on whether the owner should approve a change request, and on the ultimate cost of the change. If the contractor was deficient in its change order management, this deficiency can result in the contractor's inability to recover costs and time impacts associated with the change.

4.7 Allocation of Responsibility for Project Management Problems to Delay and Damages Calculations

Various project management failures and how to determine which party is responsible for them have been discussed above. The next challenge is to show how the delay and cost impacts of these problems can be allocated to the parties and quantified.

4.7.1 Delay

Project management failures can impact scheduled project activities. A contractor's failure in procurement management, for instance, may lead to late award of subcontracts.

[24] This is a subset of Project Integration Management as shown in the PMBOK® Guide.

Consider the following as-planned project schedule excerpt and the situation described in Table 4-4 below:

Table 4-4: Sample As-Planned Schedule for Subcontract Award

Activity ID	Title	Early Start	Early Finish
00000060	CIVIL SUBCONTRACT–ISSUE REQUISITION	4/2/19	4/23/19
00000061	CIVIL SUBCONTRACT–OWNER APPROVE REQUISITION	4/24/19	5/1/19
00000065	CIVIL SUBCONTRACT–BIDDING & AWARD	5/2/19	5/23/19
00000070	CIVIL SUBCONTRACT–MOBILIZATION	5/24/19	4/28/19

- In executing its work, the contractor developed a civil subcontract requisition, or bid package, using the wrong specifications for earthwork and road construction. In this case, the earthwork specification included an incorrect statement of the scope of the road work.
- The owner, toward the end of his allotted review period, discovered the error and sent the requisition back to the contractor to be corrected.
- The contractor did not have any float in his schedule for rework of his documents, so the bidding of the civil subcontract was delayed.

If a schedule delay analysis then showed that this activity was on the critical path of the project and there was no concurrent delay, the responsibility for the delay would lie with the contractor. The actual number of days of delay attributable to the contractor's failure to prepare a proper bid package would then be evident.

Table 4-5 shows an example of the impact of a contractor's failure to properly execute his testing and turnover plan.

Table 4-5: Contractor's Planned Schedule for Walkdown, Inspection, and Punch List

Activity ID	Activity Description	Early Start	Early Finish
E01240120	SOLID REACTOR SE-C-05–TESTS AND PRE-COMMISSION	8/31/19	9/14/19
E01240200	SOLID REACTOR SE-C-05–FUNCTIONAL CHECK-OUT	10/8/19	10/9/19
E01240510	SYSTEM PACKAGE AVAILABLE TO OWNER	7/19/19	7/19/19
E01240520	OWNER PRE-WALKDOWN (STARTUP TEAM)	7/30/19	7/31/19
E01240530	JOINT WALKDOWN (CONT/OWNER) INSPECTION/PUNCH LIST SYSTEM	9/15/19	9/18/19
E01240540	CONTRACTOR COMPLETE PUNCH LIST ITEMS	9/19/19	9/28/19
E01240550	CONTRACTOR CHECK SYSTEM–QC DOCUMENTATION	9/29/19	10/2/19
E01240560	SUBMIT QC DOCS TO OWNER STARTUP TEAM	10/3/19	10/3/19
E01240570	FINAL JOINT WALKDOWN/AGREEMENT ON EXCEPTIONS	10/4/19	10/5/19
E01240580	SYSTEM ACCEPTANCE BY OWNER	10/6/19	10/6/19
E01240590	OWNER APPROVES PACKAGE	10/7/19	10/7/19
E01240625	SOLID REACTOR SE-C-05–TURNOVER	—	10/7/19

In this case, the contractor developed and submitted a testing, pre-commissioning, and turnover plan to the owner that was approved prior to the start of the actual field work. Plan in hand, the owner showed up on time for the first "joint walkdown" of the reactor on September fifteenth and asked for the design documents that were supposed to be provided at the walkdown.

Unfortunately, the contractor failed to provide the updated process and instrumentation diagrams and other design documents that were necessary to guide the inspection and punch listing of the work. The drawings had not been completed. Two weeks later, the contractor produced the drawings, and the joint walkdown commenced. This resulted in a two-week delay in the turnover of the facility to the owner.

Again, if a schedule delay analysis then showed that this activity was on the critical path of the project and there was no concurrent delay, the responsibility for the delay would lie with the contractor.

4.7.2 Costs and the Allocation of Damages

Regardless of who is the initiator, a claim must contain three essential elements before any damages can be recovered:

1. **Entitlement**. What actions and/or inactions by the other party or parties to the contract resulted in compensable liabilities?
2. **Cause and Effect.** Do the entitlements for the problems encountered bear a direct relationship to the financial damages that were incurred?
3. **Damage Computation**. What methodology or methodologies will be used for presenting and pricing the compensable (recoverable) damages?

There are numerous entitlements that are recognized by the U.S. federal courts, state courts, and arbitration panels leading to the recovery of damages related to a construction claim. These entitlements include:

1. Acceleration
2. Acts of God/Weather
3. Cardinal Change (Breach of Contract)
4. Constructive Change
5. Defective and Deficient Contract Documents
6. Delays
7. Differing Site Conditions
8. Directed Change
9. Disruption
10. Implied Warranty
11. Impossibility of Performance
12. Maladministration
13. Owner-Furnished Items
14. Strikes
15. Superior Knowledge
16. Suspension
17. Termination
18. Unjust Enrichment
19. Variations in Quantities

4.7.2.1 The Cause-Effect Relationship

Cause-and-effect relationships linking the entitlements to the damages are most often found in the project's contemporaneous documentation, *i.e.*, updates of the project plans and schedules, periodic reports, correspondence, change orders, requests for information, meeting minutes, superintendents' logs, etc., as discussed above.

The ability to quantify damages, however, depends on the availability and analysis of detailed and verifiable man-hour and cost data. Having demonstrated the nature of the loss and the connection between the loss and the other party's actions and/or inactions, the success of the claim will often hinge on how well the claimant can calculate and prove its damages. Conversely, the parties to a contract should understand the principles of proof of damages to properly and fairly evaluate a claim, grant the cost and time request if they are justified, or reject the claim if convincing and supportable evidence is not presented.

To demonstrate the cause-effect link of the issue entitlements to damages, development of a Cause-Effect Matrix can be helpful. To start with, a simple example of the cause-effect relationship is provided:

CAUSE:

The contractor produced defective design specifications.

EFFECT:

The hydrostatic load tests performed in the field failed. The project will be delayed.

This simple concept is shown graphically in Figure 4-1.

Figure 4-1: Sample Cause-Effect Matrix 1

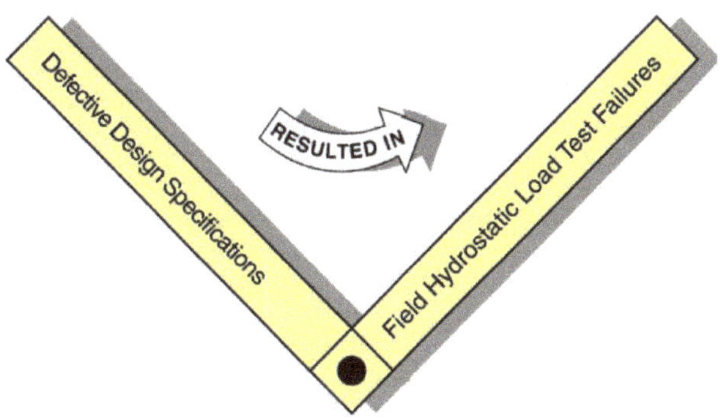

Of course, as causes and effects become increasingly complex, say with primary and secondary causes and intermediate effects, the cause-effect matrix becomes more complex, as shown in Figure 4-2.

Figure 4-2: Sample Cause-Effect Matrix 2

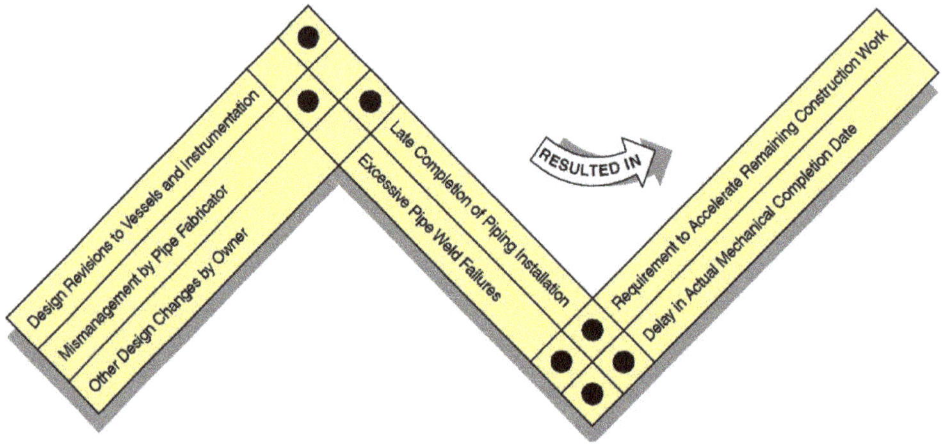

Ultimately, a Cause-Effect Matrix for a contractor's delay and disruption claim that can show the link all the way from primary causes, through intermediate effects, to the amounts of damages allocated to each party becomes quite complex, as shown in Figure 4-3 on the next page.

4.7.2.2 Cost Allocation Example

After a cause-effect link is established, the analyst will attempt to allocate costs associated with project management problems to specific cost accounts. The "attempt" qualifier is used because cost records that can make the association are not always kept. When adequate cost records are kept, however, the analysis is relatively straightforward.

Using the example of delay to the turnover of a reactor shown above, the contractor kept records of its direct labor associated with that facility. A mechanical crew was assigned to the facility to be used for punch list work but was used only minimally while waiting for the results of the joint owner/contractor walkdown. This resulted in increased labor costs for that operation compared to the contractor's budget.

When the work was finally completed, the contractor claimed for increased costs on the facility and blamed them on overzealous inspection and contrived delay by the owner. Overzealous inspection notwithstanding, the contractor failed to take responsibility for the additional man-hours it caused and inappropriately applied the entirety of its overrun to a change request and claim.

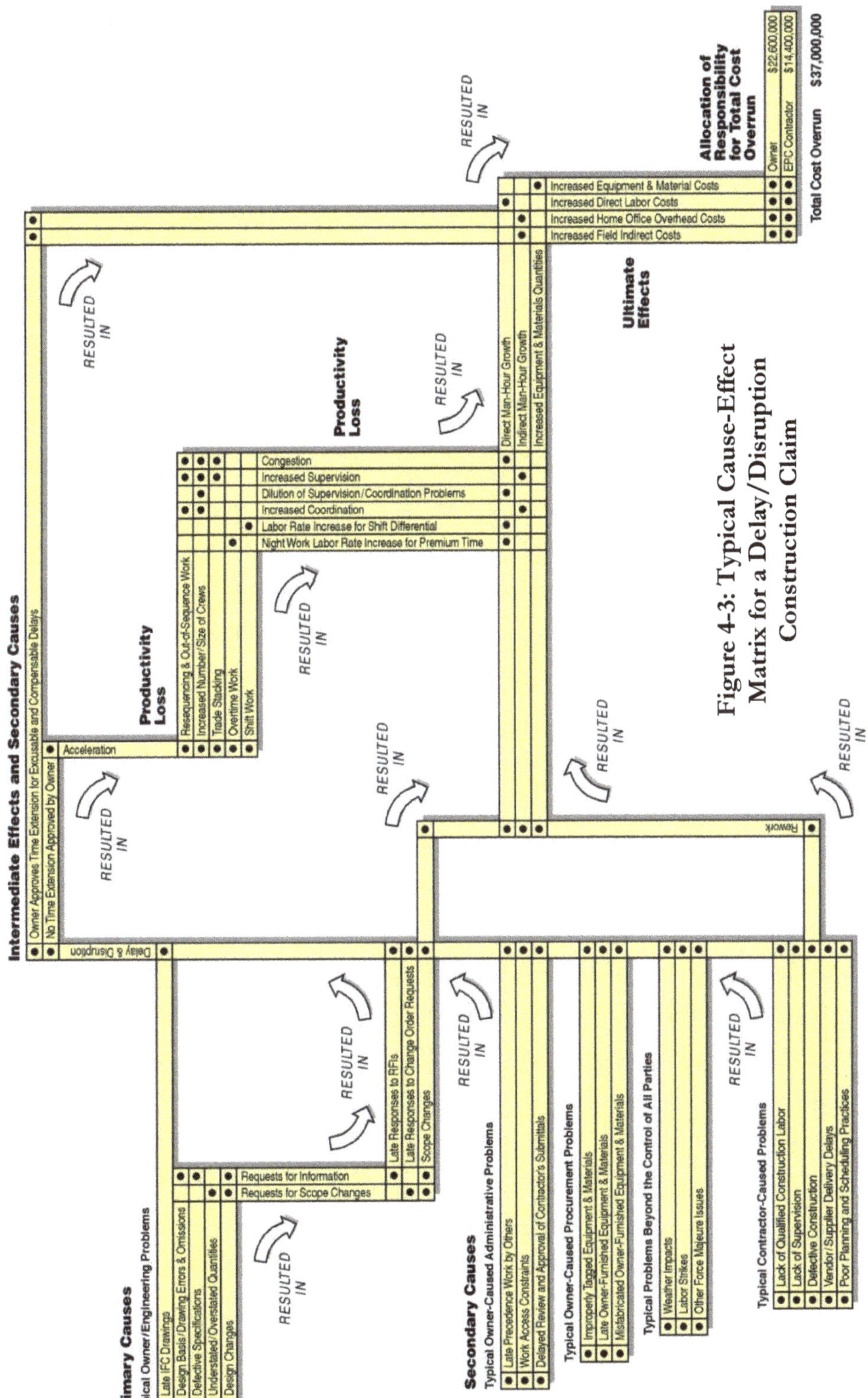

Figure 4-3: Typical Cause-Effect Matrix for a Delay/Disruption Construction Claim

From the contractor's job cost records, Table 4-6 is an example of the actual man-hour costs charged to the facility for the two-week period when the contractor's mechanical crew was underutilized by the delay to the walkdown.

Table 4-6: Contractor's Job Cost Records

Cost Code	Labor Cost	Labor Hours	Equip Cost	Supply Cost	Mater Cost	Subs Cost	Cost Code Description
922543700	$3,647.38	129.5	0	0	0	0	Turnover Reactor SE-C-05–Mechanical
922643700	$84.47	2.5	0	0	0	0	Turnover Reactor SE-C-05–Instruments
922943700	$2,615.73	108	0	0	0	0	Turnover Reactor SE-C-05–Pipe Fitting
Total	**$6,347.58**	**240.0**	**0**	**0**	**0**	**0**	—

The costs attributable to the contractor's failure to properly execute its turnover plan could then be deducted from its claim.

4.8 Examples of Project Management Problems from Actual Cases and How They Were Analyzed

The following additional examples of project management problems are discussed below.

- Failure to Produce the Project Schedule on Time
- Failure to Provide Commissioning Procedures
- Building from the Wrong Drawings

4.8.1 Failure to Produce the Project Schedule on Time

An EPC contractor failed to produce a reliable project schedule in the time required by the contract. The contractor blamed its failure to do so on everything but its own negligence, *e.g.*, the owner was too picky, it did not get the Notice to Proceed when the contractor wanted it, and the ground was frozen, so it had to wait for the ground to thaw.

The contract provided a clear requirement for type, level of detail, and date required for development of the project schedule. The owner pointed out the contractor's deficiencies by responding to several of the contractor's submittals for approval, and ultimately, the schedule was approved late. The contractor blamed this delay on the owner, but a review of contemporaneous correspondence showed that the contractor was at fault. For example:

- Correspondence showed that the contractor's own scheduler admitted that the schedule contained "inappropriate items." In this case, this deficiency meant that the scope was not entirely covered and that some activities could not be completed in the time allotted.
- The contractor's Project Manager admitted in an email that it was "faking" the schedule, *i.e.*, the submittals were not based on full knowledge of what had to be done and how long it would take to perform the work.
- In an email, the scheduler admitted that the contractor could not achieve planned production with available manpower.
- The Project Manager admitted in an email that his Superintendent ignored the schedule, refused to provide progress updates, and "*did planning in his head.*"

As a result of these findings, the contractor's claim of the owner delaying approval of its schedule was discredited, and the associated costs were not recoverable.

4.8.2 Failure to Provide Commissioning Procedures

An EPC contractor failed to provide complete commissioning procedures in the time required by the contract. The contractor blamed this failure on the owner's overzealous review of the procedures and its failure to approve them. Contemporaneous documents showed a different story, however:

- A written notice from the owner prior to any submittals pointed out that the contractor's failure to provide the needed commissioning procedures on time was going to result in a two-week delay to the start of commissioning.
- Additional owner correspondence pointed out that the commissioning sequences, when finally submitted for review, were impossible to execute because they did not consider the plant shutdown schedule. This resulted in further delay for rework of the procedures.

The contractor's delay claim was discredited, and it was unable to collect any of the additional commissioning costs for the two-week period attributed to its failure to provide the proper procedures in a timely manner.

4.8.3 Building from the Wrong Drawings

An EPC contractor's construction crew executed its work from drawings that were not "Approved for Construction." This approval process was required according to the engineering procedures developed under the contract. The drawings contained errors. The contractor had been warned of potential problems with the use of these drawings but persisted in using them. This eagerness to use the drawings was due to the contractor's field supervisors' frustration with the drawings arriving late from its engineering department. Contemporaneous documents showed that the contractor was at fault. Examples include:

- Correspondence from the owner showed that it warned the contractor on three occasions that workers were building foundations from drawings marked "Preliminary" and that these drawings were incorrect.
- The use of unapproved and incorrect electrical conduit drawings resulted in the casting of incorrect foundation penetrations into a substation building.
- The superintendent's log recorded that pipe rack drawings were changed in the field and the racks were constructed without engineering review. This resulted in the need to reengineer and reconstruct some of the pipelines on the pipe racks because the racks would not accommodate expansion loading in their as-built condition.

As a result of these findings, the contractor's claim for additional costs for rework and reengineering on the above work was denied.

4.9 Summary

The application of sound project management practices provides construction project stakeholders with the means to meet their objectives. When sound project management is lacking, it leads to delays and extra costs for all parties, typically resulting in a dispute of some consequence.

The application of sound project management principles on a project is not only the means to the project's ends, but, during disputes, can also provide a level of defense against claims for delay and additional costs that were based on alleged management deficiencies. Demonstrating that you used sound project management practices, and that the other party did not, can help your side win disputes.

4.9 – Summary

A systematic and supportable methodology should be employed to assess project management problems caused by either of the contracting parties. This methodology includes:

- Analysis of project management requirements stated in the contract;
- Analysis of the clarity of those requirements;
- Comparison of the project management requirements to those generally accepted in the construction industry, such as PMI and CII;
- Analysis of contemporaneous project documentation;
- Evaluation of the parties' conformance to the project management requirements of the contract and to those generally accepted in the construction industry;
- Development of cause-and-effect relationships between the project problems and the ensuing damages; and
- Allocation of the responsibility for project management problems to delay and damages calculations.

Chapter 5

Typical Problems Leading to Delays, Cost Overruns, and Claims on Process Plant and Offshore Oil and Gas Projects

Chapter 5

Typical Problems Leading to Delays, Cost Overruns, and Claims on Process Plant and Offshore Oil and Gas Projects

Owners and contractors face enormous challenges to complete billion-dollar engineering and construction projects for process plant and offshore oil and gas facilities. The problems that often occur can lead to significant delays, cost overruns, and claims. These typical problems include, but are not limited to, the following:

1. Insufficiently defined Front End Engineering Design (FEED);
2. Inadequate design basis for production rates and properties;
3. Inaccurate contractor cost estimates;
4. Ambiguity of the contract documents;
5. Inadequate documentation;
6. Multiple change orders;
7. Insufficient management of contractor design and construction interfaces;
8. Insufficient and inexperienced owner technical personnel;
9. Inadequate baseline schedule development and updating by contractors;
10. Insufficient and unreliable integrated master project schedules;
11. Insufficiently sized camp facilities for housing the on-site construction and startup work force on remotely located projects, leading to delays and large cost increases for additional camp construction or hostels;
12. For offshore oil and gas projects, incomplete onshore fabrication prior to shipping leading to large amounts of carryover work offshore; and
13. Failure by owners to have a sufficient and experienced project management team in place to manage change orders, requests for time extensions, and claims.

This section describes these problems and presents recommendations to minimize the potential that they will occur and, if they do occur, to mitigate their impacts.

5.1 Insufficiently Defined FEED, Causing Change Orders, Delays, and Cost Overruns

As one of the stage-gate steps in the development of a project, the owner typically contracts with an engineering contractor to prepare a FEED package to further the development of its conceptual design, which was previously prepared during an earlier stage of the project development cycle. For process plant projects such as refineries, chemical and petrochemical plant projects, oil and gas facilities, gas plants, and liquefied natural gas (LNG) projects, the FEED package will typically include a process description and design basis description, input and product production and consumption rates, an optimized heat and mass balance, process flow diagrams (PFDs), partially developed piping and instrumentation diagrams (P&IDs), equipment process datasheets and specifications, instrument datasheets and specifications, an equipment list, a line list, utility flow distribution drawings, a preliminary plot layout, environmental design requirements, and perhaps other basic design information.

The FEED package design information is not a final design and, therefore, is not sufficient to procure all equipment and materials and construct the project. However, if the design requirements are properly defined, an experienced engineering and construction contractor can use the FEED package to develop a lump-sum bid for the EPC phase of the project. To do that, the EPC contractor, during the bid phase, must further develop the design such that it can estimate quantities, equipment, materials, and other requirements for the project, adding appropriate contingencies for design development.

Because the FEED package is not a completely developed design package with final drawings and quantity takeoffs sufficient to procure equipment and materials and define the construction packages, the EPC contractor may underestimate its bid costs and the time required to perform the work. If the owner selects the lowest bid, that EPC contractor may eventually learn that it either left out scope, did not provide sufficient contingencies in its quantity estimates, or must modify the design to a greater degree than it planned so that it can comply with the detailed specifications that are required for the project.

The EPC contractor then submits numerous change orders to attempt to recover for these problems, leading to delays in completion of the project and increased costs. Owners often consider that many of these design problems are not compensable because they believe that the EPC contractor's scope is to further develop the design, and thus the modifications that the EPC contractor makes to the design documents are the result of this design development. However, the EPC contractor may view differences between the preliminary drawings and other design documents, and the equipment, instrumentation, and other specifications in the FEED as design conflicts or inaccuracies for which the owner is responsible and which will require increased costs and time to resolve. Thus, owners are faced with the dilemma of either resolving the FEED package design issues that are identified during the EPC phase by agreeing to change orders that include additional costs and time extensions or face large delay and disruption claims at the end of the project.

RECOMMENDATION: The owner should conduct a cold eyes review of the bid package and FEED documents that it intends to submit for obtaining lump-sum bids for the EPC phase to ensure that the FEED package and detailed design requirements are well-defined, the requirements are clearly stated and unambiguous, and the contract documents and project procedures are properly drafted for a lump-sum contract. Also, owners should consider including in the EPC Phase a 30 to 60-day period for the successful EPC bidder to further review the FEED package and requirements in detail and advise the owner of any discovered problems so that these problems can be resolved before too much time is lost. Owners may also consider including in the EPC contract a provision to consider change orders to resolve design conflicts or ambiguities during this initial 30 to 60-day period. The contract should also contain an order of precedence provision that states which documents control the detailed design requirements should conflicts and ambiguities exist in the FEED package documents.

5.2 Inadequate Design Basis for Oil, Gas, Chemical, and Petrochemical Production Rates and Properties, Leading to Design Changes that Delay the Project and Increase Its Costs

Owners often attempt to improve the timing of project delivery by identifying the design basis of the FEED work using preliminary oil, gas, chemical, and petrochemical properties, compositions, and flow rates. The thinking is that the completed FEED package can be updated with the updated design information, and the detailed design contractor can update the design with the final feed and product properties, compositions, and flow rates. The problem is that the EPC contractor submitted its bid price and schedule based on the FEED package and its design basis. If the final feed and product properties, compositions, and flow rates are sufficiently different from the FEED design basis, the EPC contractor will submit change orders with delays and cost increases to resolve those differences.

RECOMMENDATION: If the owner learns that the oil and gas properties, compositions, and flow rates are sufficiently different from the FEED design basis, it should consider delaying the issue of its bid documents to the EPC contract bidders and going back to the FEED engineering contractor to update the FEED package with the new information. That will be a less costly solution compared to making those changes under a lump-sum EPC contract. It may also reduce the overall time to complete the EPC project.

5.3 Inaccurate Cost Estimates by the Contractor due to Inadequate FEED Definition Resulting in Huge Cost Overruns When Scope Definition Becomes Better Defined

In an effort to initiate the EPC phase of a project as quickly as possible, owners may prematurely send out their FEED packages to engineering and construction contractors seeking lump-sum bids. Unfortunately, consequences of such decisions can be one or more of the following:

- Highly qualified EPC firms will decline to bid, leaving the owner without confidence that its project can successfully proceed.
- Highly qualified EPC firms will submit bids but will include large contingencies that may make the project uneconomical.
- Highly qualified EPC firms will recommend that the owner convert its contract basis from lump-sum to reimbursable in an effort to reduce the EPC contractor's risk but reduce the owner's confidence as to the true capital cost of the project.
- Lower-tier contractors who are eager to enter a new market will submit lump-sum bids. If they are successful, problems with the FEED documents will eventually be identified, and large change orders, cost overruns, and delays will result during the execution of the project. If the owner refuses to accept the requested change orders, the contractor may default rather than risk financial ruin.

RECOMMENDATION: The owner has few good options until it prepares a FEED package with sufficient definition to enable the EPC contractor to understand the full scope of the project. The plot plan for the process units needs to be well planned. Process flow diagrams, P&IDs, and equipment specifications need to be taken to a level of completeness that will enable the EPC contractor to obtain competitive bids for equipment, understand the civil earth-moving requirements, and quantify concrete, structural steel, piping, electrical, and instrumentation material requirements without significant contingencies. Experienced EPC contractors understand that some level of design development occurs when a FEED package is developed into "Issued for Construction" drawings and specifications. Design development is part of the EPC contractor's scope of work, and the bid price should include the cost of this engineering. While some level of scope changes typically will occur even with a well-developed FEED package, if the amount of changes exceeds six to eight percent, then the owner may be exposed to a cumulative impact claim at the end of the project.

5.4 Ambiguity of the Contract Documents, Leading to Disputes over Procedures for Progress Measurement and Reporting, Technical Specification Requirements, Cost and Schedule Control, and Change Order Management

Contract documents and cost, schedule, change management, and progress reporting procedures are often taken from previous projects, dusted off by a new project team, and assembled for use on the next project. It is unlikely that all requirements from a previous project will be appropriate for the next project, and the boilerplate needs careful review to ensure that the contractual and technical specification requirements are correct and that the progress measurement and reporting, cost and schedule control, and change order procedures are coordinated and appropriate for the type of contract envisioned for the engineering, procurement, and construction phase of the project. For example, a lump-sum contractor will be submitting less detailed actual budget and man-hour breakdowns as part of its progress reporting than it would under a cost-reimbursable contract. The level of support from the contractor that will be required for pay applications will be different for lump-sum contracts than it will for reimbursable contracts. The basis for calculating the progress and percent completion amounts needs to be defined. The owner must also decide what it will do to verify the contractor's progress and percent complete submittals. The contractor's actual labor productivity may not be discernable from the data it submits to the owner under a lump-sum contract; therefore, actual man-hours may not correlate well to percent complete. Problems also occur when the bid package does not define the supporting schedule delay calculations and documentation that the contractor is required to submit to justify a time extension.

RECOMMENDATION: The owner should have an experienced contract administration/construction claims expert review its contract procedures to ensure they are appropriate for the type of contract envisioned and are sufficiently detailed and coordinated to obtain information to evaluate the basis for progress, cost and man-hour updates, schedule updates, and productivity measurement.

5.5 Inadequate Documentation to Support Positions and Resolve Issues that Develop During the Execution of the Project or Retrospectively When Large Claims Develop at the End of the Project

During a project, contractors often attempt to submit minimal information with which an owner can evaluate the contractor's schedule progress, labor utilization, cost expenditures, and labor productivity. Issues and problems affecting cost and schedule invariably arise, and contractors may attempt to cast blame for those issues and problems on the owner and prepare requests for time extensions and changes in scope. Often, the owner may have responsibility for these issues and problems. However, without a complete historical record and documentation from the contractor's files, the owner may not have all facts. Daily reports prepared by the EPC contractor and its subcontractors, inspection reports, and RFIs between the EPC contractor and its subcontractors often tell a different story as to the real causes of problems and their effect on project performance. Also, this documentation could reveal concurrent delays for which the contractor is responsible, and which could mitigate the owner's responsibility for the cost of certain issues and problems.

RECOMMENDATION: In its contract with the EPC contractor, the owner must make a clear requirement of the contemporaneous documentation that the EPC contractor must submit to the owner during the project or include in the project database to which the owner will have unfiltered access. It is much easier to obtain documentation from the contractor if it is specified in the contract than to request this information only after issues and problems arise. Therefore, owners should anticipate the need for the contractor's documentation in the contract requirements and establish a procedure to obtain such information in the normal course of contract administration and not only after problems and issues arise.

5.6 Multiple Change Orders Are Approved During the Project or Remain Unresolved Until the End of the Project, Leading to Large Delay and Cumulative Impact Claims for Recovery of Lost Productivity Costs

Contractors may sign off on early change orders without including extra costs for productivity loss and cumulative impact from multiple changes. However, after numerous changes have occurred, contractors who were lax on recognizing the potential for such costs will notify the owner that they reserve their right to submit a cumulative impact claim to recover additional costs beyond the costs that were previously agreed to in prior change orders. Owners resist such tactics, and disputes may occur that are not resolved until after the project is completed, and often in the courtroom or arbitration hearing.

RECOMMENDATION: In their contracts, owners should include proforma change order forms that include full accord and satisfaction language that makes clear that the agreed-upon price of the change is to include any loss of productivity, delay, cumulative impacts, acceleration, and all other direct and indirect costs that arise from the changed work and the effect on the unchanged work. With this information, the contractor is clearly on notice that cumulative impact claims at the end of the project will not be considered valid. However, loss of productivity resulting from the cumulative impact of changes may be considered indirect disruption that is not foreseeable. Thus, even with contract language stating that the change order costs are to include cumulative impact costs, a contractor may still attempt to argue that such impacts were not foreseeable and, therefore, when they became known, it has entitlement to additional compensation for the costs of such impacts.

Often, the owner will not accept cumulative impact productivity loss man-hours and costs in the change orders and will require the contractor to sign the change orders without such costs before it receives payment for the additional or changed work. If this situation occurs, then the contractor should include in its transmittal letter with the change order, or in a separate letter, a reservation of its right to later make a claim for such productivity loss.

Owners should also require that the change order documentation from the contractor include a direct and indirect man-hour breakdown by discipline. Cumulative impact claims are usually evaluated by comparing the change order man-hours to the base scope man-hours. Without the change order man-hours, the owner does not have all information that it may need to perform its own evaluation of potential claims for the cumulative impact of multiple changes.

Absent contemporaneous productivity data, *i.e.*, contemporaneous man-hour and installed quantity data, the contractor will not have the best data to support a productivity loss claim that results from the cumulative impact of multiple change orders or other causes of productivity loss. A measured mile analysis of the contractor's productivity data is often considered the best type of analysis to support a productivity loss claim. This analysis compares the contractor's productivity during an unimpacted period to the contractor's productivity when the owner-caused impacts occurred. The difference represents the productivity loss caused by the impacts. Therefore, the contractor's management should require its field personnel to collect man-hour and installed quantity data on a weekly basis throughout the duration of the project.

5.7 Insufficient Management of Multiple Prime Contractor Design and Construction Interfaces, Leading to Design Changes, Delays, and Cost Overruns

Owners often place responsibility on multiple prime contractors to coordinate and integrate their design and construction work to avoid interferences and delays. While such contractual language is commonly used, one contractor cannot require another contractor to do anything that may alter its design or method of construction performance. If such consequences occur, the affected contractor(s) will seek recovery from the owner for its time and cost impacts of such coordination and integration, and any changes that result from ambiguous design and/or construction responsibilities that are assigned to the various contractors.

RECOMMENDATION: The owner should be proactive when it receives the detailed design documents, work plans, and schedules from its various prime contractors and attempt to anticipate such interface problems. The owner's project management team should be an active participant in coordination meetings between contractors and attempt to amicably resolve problems rather than relying on contract language that states that the contractors must coordinate and resolve problems. An overall engineering, procurement, and construction management (EPCM) contractor may be a better strategy if the owner's project management team is stretched too thin to accomplish the identification and control of interfaces between multiple prime contracts.

5.8 Owners Often Do Not Have a Sufficient Number of Experienced Personnel to Manage the Technical Aspects of a Project, Leading to Delays in Reviews of Design and Procurement Packages, and Late Responses to Requests for Information from the Contractor

Contractor delay and disruption claims often arise because the owner does not respond in a timely manner to documents submitted by the contractor for owner review and approval, or to RFIs requiring clear responses and direction.

RECOMMENDATION: Owners need to ensure that their technical and project management teams are adequately staffed to review documentation and provide adequate responses in a timely manner. Requiring the contractor to include all owner review activities in its schedules will enable the owner to better anticipate the timing of such requirements. The owner should also require that no owner-responsible activities are within a reasonable number of work days (ten would be a good number) of the critical path on the baseline schedule, and then the owner should closely monitor those owner-responsible activities as the schedule is updated to be aware if such activities become critical as the work progresses. As for RFI responses, the owner should conduct a cold eyes review of the design package before the contract is let to avoid problems with the information provided to the contractor. The contract procedures should also define a reasonable period for the owner to review RFIs and provide an adequate response and direction to the contractor. Also, the owner could utilize a review period up front in the contract schedule such that the contractor must have all of its information requests submitted during this time period or waive its right to time extensions if later requests are made that could have been made in the earlier period.

Another strategy that the owner could employ is to have contract language that defines the number of days that the owner has available to review and submit comments on the contractor's submittal packages, after which the contractor's submittals are deemed to be approved and the contractor is entitled to proceed with the follow-on work. Also, the owner should include a contractual provision that limits the number of submittals (drawings, specifications, purchase orders, etc.) that the contractor can submit during any given period to prevent overloading the owner's review team.

5.9 Inadequate Baseline Schedule Development and Updating by Contractors, Leading to Unreliable Progress Measurement, Uncertain Critical Paths, and Inability to Mitigate Delays

If the contractor's baseline schedule: 1) does not adequately define the period of performance for all scope requirements; 2) contains logic errors and excessive constraints; or 3) is inconsistent with its bid basis, *i.e.*, activity durations do not match the man-hours determined by the contractor's bid estimate to perform a given activity, the baseline schedule will be flawed and not useful to measure the contractor's work progress. The baseline schedule problems could carry over to the schedule updates, and the schedule problems could be compounded by insertion of incorrect progress and/or inaccurate actual start and actual finish dates for activities. Also, an inadequate baseline schedule and schedule updates will not provide a valid basis for determining if critical path delays that may entitle the contractor to a time extension or compensable delay have occurred. Likewise, an inadequate baseline schedule and schedule updates will not provide the owner an adequate basis to determine if acceleration should be performed on a paid basis or provide a proper basis to determine if the owner is entitled to liquidated damages after considering the contractor's entitlement to a time extension.

RECOMMENDATION: Owners should consider a review of the quality of the contractor's baseline schedule and periodic schedule updates by an experienced forensic schedule delay analyst to determine if problems exist. Items of concern include the following:

- Does the schedule include activities for all scope requirements in the contract?
- Are there excessive activity durations, indicating that insufficient detail has been provided in the schedule to enable accurate measurement of progress?
- Does excessive float exist on activities, indicating optimization of the schedule logic may be appropriate?

- Do excessive negative lags exist between activities, suggesting a change from a finish-to-start relationship that creates the negative lag to a start-to-start relationship between the activities or a change in the link to a different activity?
- Is there a significant difference in the planned versus actual labor resources, indicating potential loss of productivity or potential problems with the bid?
- Is there a large difference in the planned duration compared to the actual duration of a significant number of activities, indicating delays have occurred or the planned basis for activity durations was in error?
- Is there a large difference in the planned lags between activities compared to the actual lags of a significant number of logic ties, indicating delays have occurred or the planned basis for lags was in error?
- Was there a change in the schedule calculation mode from "retained logic" to "progress override" or vice versa, indicating that the basis for forecasting the date of the project completion activity has changed, which date may be unreliable?
- Is there an excessive use of constraints, such as "start-no-later-than" or "must-finish-on" constraints, and has the reasoning for these constraints been examined and justified? An overuse of constraints may cause an inappropriate calculation of the critical path for completion of the project.
- Do all activities have at least one successor (except the final activity) and at least one predecessor (except the first activity)?
- Have new activities or increased durations of existing activities been included for change orders?
- Were the engineering percent complete value and quantities of material received at the start of field construction or major fabrication checked to be sure that the progress is consistent with the contract basis for progress measurement?
- Do any activities have large negative or positive lags in the relationships? Activities with large positive and negative lags should be identified, as they can distort the logic. Can large positive lags be better represented by adding new activities?
- Do any activities have progress but no actual start date?
- Are any activities completed without an actual finish date?

5.10 Owners Do Not Develop a Sufficient and Reliable Integrated Master Project Schedule that Results from Pulling Together the Work Activities from Multiple Prime Contractors

Inadequately prepared contractor schedules, if integrated "as-is" into an overall Master Project Schedule, may provide false calculations of the overall critical path of the entire project. Also, if the activity interfaces among various contractors are not properly identified, delays by one contractor may not properly show the resulting impacts to other contractors. Thus, delays may go unrecognized and mitigation options are diminished until corrections to the Master Project Schedule are made.

RECOMMENDATION: Owners should first determine if the schedules that are developed by the prime contractors are properly prepared and devoid of schedule logic or critical path calculation problems as identified in Section 5.9 above. Then, a thorough review of the need for activity ties from one contractor's activities to the activities of other contractors should be performed, and additional relationships should be added, where necessary, to fully integrate the schedule.

5.11 Insufficiently Sized Camp Facilities for Housing the On-Site Construction Work Force and the Owner's Startup and Initial Operations Personnel on Remotely Located Projects, Leading to Delays and Large Cost Increases for Additional Camp Construction or Hostels

On remotely located projects, the size of the camp that is needed to house the construction direct and indirect labor required to build a project may be understated. The camp sizing is often determined based on a premature estimate of the required labor to complete all work activities required to complete the project. Labor force sizes are often underestimated because of premature estimates of the quantities of materials required for the project, such as civil earthwork, and quantities of concrete, structural steel, piping, electrical, and instrumentation. If the quantities increase significantly after the camp size is established, then more construction man-hours will be required.

Also, unrealistic assumptions regarding construction labor productivity can result in the need for significantly more construction labor man-hours or more workers to complete the project by the established completion date. Increased labor requirements can then lead to additional loss of productivity and increased costs as a result of crowding, trade stacking, dilution of supervision, and acceleration measures to attempt to maintain the completion date. If the owner's startup and operations team arrives at the project site when it was originally planned, additional beds may be required if the construction has been delayed or additional labor was added to perform the project work. This too becomes a problem if the camp is too small to handle these additional personnel at the project site. Options to mitigate this problem include bringing in hostels or building additional housing structures to provide more bed space, working longer hours and/or multiple shifts to perform the increased work scope or delaying the completion of the project.

RECOMMENDATION: Owners should avoid a premature establishment of the camp size requirements for a project until the equipment and materials that will be required for the project can be accurately quantified. Also, owners should scrutinize the contractor's basis for estimating construction labor productivity. What was the actual labor productivity on prior projects in the area where the project will be built? Will imported labor from other countries be required? What is the direct versus indirect labor ratio assumed by the contractor for the entire site work force, and is this realistic based on the qualifications and experience of the expected labor planned to be used to perform the project work? Is the contractor's CPM schedule resource loaded with the labor man-hours resulting from the most accurate contractor estimate for the labor requirements needed for the project? Is the contractor's CPM schedule accurately prepared so that the peak labor needed for not only the early start dates but also the late start dates indicated in the schedule can be determined and used as a basis for camp size? All these factors will directly affect the size of the camp needed for the project and should be thoroughly vetted to ensure that the peak labor work force can be accommodated.

5.12 For Offshore Oil and Gas Projects, Incomplete Onshore Fabrication Prior to Shipping, Leading to Large Amounts of Carryover Work Offshore at Much Higher Costs

Owners and their contractors often decide to "sailaway" the components of their offshore oil and gas projects being fabricated and assembled onshore in various fabrication yards before all required work on these components is completed. These larger components include topsides decks and hulls, which weigh hundreds of tons or more. Enormous transport vessels are needed to carry these large components to their offshore installation locations. Decisions to complete the work offshore are often made to avoid missing the window of availability of the limited number of large transport vessels. While the cost of completing the work offshore is more expensive than it would be onshore, the cost of missing a transport window can include large demurrage charges or delay to the project while waiting for another transport vessel to become available.

RECOMMENDATION: Owners should carefully review the accuracy of the contractors' progress reports and work to complete forecasts in their schedules and progress payment requests. Also, the contractor's schedules should be evaluated to determine if they are properly prepared and void of schedule logic or critical path calculation problems as identified in Section 5.9 above. Significant change order requests by the contractor, large changes in the required quantities of installed equipment and materials, large weight changes of the topsides decks or hull, and significant decreases in the contractor's labor productivity are indications that the completion date may slip, carryover work may be necessary, and plans may need to be prepared to complete the fabrication work offshore.

5.13 Failure by Owners to Have a Sufficient and Experienced Management Team in Place to Manage Change Orders, Requests for Time Extensions, and Claims

Owners often do not have sufficiently qualified and experienced engineering, construction, and project management personnel to manage a large EPC project. They may, however, proceed with a project thinking that: 1) the lump-sum EPC contractor will properly perform all the necessary functions, 2) they will select an EPCM contractor to act as their agent to execute the contract and hire individual specialty contractors to perform the construction work, or 3) they will seek to bring in a team of experienced personnel who come from other corporate cultures but have not previously worked together. This owner's project management team is now charged with managing the lump-sum EPC contractor but must first develop the procedures to do so. After the initial good feelings at the beginning of a project are over, contractors often become aggressive and submit change orders and claims if they feel entitled to additional cost recovery. The owner's project management team then becomes upset with the contractor, and the project relationship suffers.

RECOMMENDATION: In any of these scenarios, the owner should implement a robust Claims Prevention Program. This Program would include having sufficiently experienced staff to deal with contract administration, delay/extension of time, scope change requests, acceleration, and other issues that could result in claims and/or arbitration/litigation if not amicably resolved. If the owner does not have such staff in-house, then the owner should consider retaining a project management consultant who has the relevant experience to deal with these issues.

Chapter 6

Claims Prevention—Early Mitigation Steps

Chapter 6

Claims Prevention—Early Mitigation Steps

6.1 Introduction

Claims prevention can be accomplished most effectively during the design phase of a construction project. When the design is sufficiently complete to enable qualified contractors to submit bids for the construction work, the careful preparation of well-considered contract documents offers the next greatest opportunity for claims prevention. Once contracts are executed and construction begins, the prevention of claims becomes more difficult, although the impact of claims can still be mitigated through timely management actions and effective contract administration. Figure 6-1, "Project Transitions," shows how the responsibility for claims control is transferred over the life of the project from the planner to the designer and then to both the construction contractor and construction manager.

Figure 6-1: Project Transitions

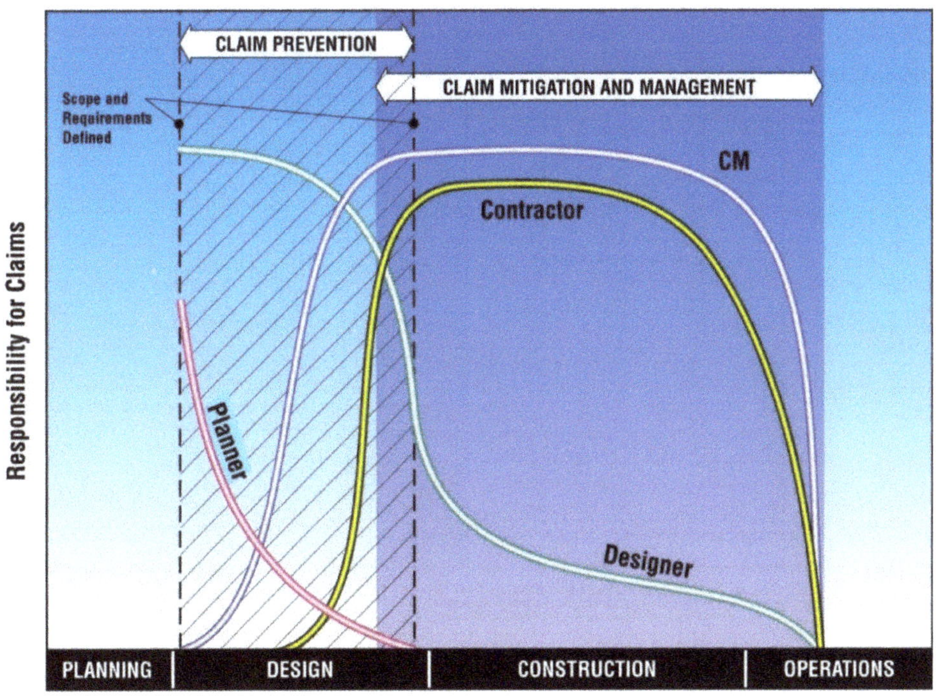

The owner remains responsible, however, for claims prevention and mitigation activities during all phases of the project.

6.2 Quality Contract Documents

The successful project manager is an effective manager of contracts. From the owner's point of view, this involves three basic steps:

- Develop contract documents that provide a clear and nonconflicting basis for a contractor to assemble a bid. The contract documents must reflect the intent of the project in language that is unambiguous and provides a basis for a contractor to plan its means, methods, and sequences. Production of plans and specifications merely to satisfy owner goals of form and function is no longer the single objective.
- Contract administration has taken on new meaning. It is not merely the enforcement of the terms and conditions but includes the use of the terms and conditions to adapt to the changing environment, ongoing contract interpretation and management of decisions necessary to keep the project moving forward. Both the contractor and the engineer have duties under the terms of the contract.
- Perhaps the most important step and the one most frequently overlooked is the importance of record keeping or documentation. Does the engineer have the hard facts to support decisions made during contract interpretations and to defend against unreasonable allegations on the part of the contractor expressed in the wording of a total cost claim? Without documentation, the engineer representing its owner is at the mercy of the courts.

Most claims are based on errors, omissions, conflicts, and ambiguities in the contract documents and/or erroneous interpretation of those documents. The engineer is in the middle of this process. No longer can claims be put off as an element to be resolved after the project is completed. The increasing exposure of the design professional makes it necessary for that design professional to take an active role in a claims prevention and mitigation program.

The design professional plays a major role in a claims prevention and mitigation program prior to award of the construction contract. After all, the documents prepared by the design professional are often used as the basis for a claim. Design professionals are not perfect. There will be errors, omissions, and conflicts. The secret of a successful claims prevention and mitigation program is to minimize those errors and omissions and then to respond positively and reasonably when an error and/or omission is uncovered.

A claims prevention review focuses on the principal causes of contractor claims. Value engineering and constructability reviews, on the other hand, focus on alternatives to reduce costs yet accomplish the same overall objectives. These alternatives typically involve means, methods, materials, form, equipment selection, etc. A claims prevention review focuses on potential errors, conflicts, omissions, ambiguities and misrepresentations both in terms of the contract documents and the system to administer, interpret and manage those contracts.

It is important to recognize what is not included in a claims prevention review. A claims prevention review does not look at better or less expensive ways to accomplish the objective. It does not address the technical adequacy of the solution. It does attempt to find and recommend elimination of potential causes of claims.

The following questions during the claims prevention review help to evaluate contract documents:

- Are the contract documents clear, complete and enforceable?
- Does the contract language use the common and normal meanings of words?
- Have the contract documents been reviewed to ensure conflicts do not exist between various sections?
- Do the contracts use exculpatory language inappropriately?
- Are the contract documents fair and reasonable?
- Do the contract documents allocate risks to the party best able to control those risks?
- Have the architectural and engineering disciplines taken sufficient precautions to ensure the design is reasonably free from errors?
- Do the contract documents adequately support the terms of payment selected, *i.e.*, fixed-price, cost-reimbursable, etc.?
- Are expectations clearly communicated?

6.3 Review and Approval of Detailed As-Planned Schedules

Creating a detailed as-planned schedule that identifies the scope of the work, the activity relationships, milestones and completion requirements is vital to the proper planning of a project. A contractor should perform the following tasks to review and check its as-planned schedules, ensure their completeness, accuracy, and reasonableness, and allow for their timely approval by the owner:

1. Verify that the schedule includes all work that must be performed.
2. Check the level of detail proposed. Is it consistent and balanced throughout the network, or is it vague in certain areas? Is the level of detail adequate to plan, schedule, coordinate, monitor, control, and report on the progress of work?
3. Check for compliance with all contract specifications related to the schedule.
4. Check to ensure that all owner-related functions outlined in the contract documents are properly incorporated. These include:
   - Access and availability dates for physical areas of the project;
   - Intermediate completion dates established for follow-on contractors;
   - Delivery of owner-furnished materials and equipment;
   - Approval of shop drawings, submittals, and samples;
   - Inspections as required;
   - Joint occupancy dates; and
   - Beneficial occupancy dates.
5. Check the project milestones and constraints established in the network and identify if they are contractual, absolute, or preferential.
6. Check if restraints in the schedule logic create incorrect critical paths.
7. Evaluate experience relative to this type of project. Spot-check relationships or work phases and their timing, *i.e.*, structural steel erected to job completion, setting of major equipment to job completion, etc.
8. When comparing the schedule for a similar job, are necessary activities included, are durations correct, and does the project duration fall within a reasonable variation range?
9. Perform a one-to-one data check to validate the consistency of the computer tabulation and the network logic if a CPM is used.
10. Clearly define the size and type of operation for each activity period.
11. Ensure that the activities are sufficiently small in duration and scope for accurate time estimation and tracking.
12. Make reasonable activity durations.
13. Properly sequence concurrent activities.
14. Review the proposed logic sequence and note any exceptions that might be taken. Validate absolute logic conditions and confirm key preferential logic conditions.

15. Spot-check activity durations for quantities involved, crew size requirements, and productivity factors. Challenge durations when appropriate.
16. Is it possible to complete each activity described in the allocated time, given the resources available?
17. Highlight the first five to seven paths of criticality (paths of least float) to review and understand the controlling logic and mathematics of the schedule. Determine if the critical path is proper and reasonable.
18. Check the plan to see if all major equipment and material restraints and delivery dates are accurately reflected.
19. Is the lead time for submittals and approvals realistic?
20. Check for involvement of subcontractors and suppliers and see if they are accurately reflected. Are dependencies clearly defined? Any critical deliveries?
21. How is weather reflected? Are there any seasonal weather restrictions to consider?
22. Prior to submitting a detailed as-planned schedule for the owner's approval, obtain approvals from an internal project management team including:
    - Project manager
    - Project controls manager
    - Operations manager

6.4 Constructability and Biddability Reviews

We have discussed the importance of a review of plans for errors, omissions and conflicts that would lead to a claim for defective construction documents. In addition, work products of the engineering department or design professionals can be subject to constructability and biddability reviews utilizing the services of the resident engineers and inspectors or outside consultants who specialize in this business.

After an owner decides to build something, the three most important parties in this process of construction are the design professional, the resident engineer team, and the contractor. If these reviews are not performed, the contractor is often the innocent victim of a contract package that is deficient in terms of constructability and biddability.

Is the contract package an adequate "basis for a bid?" Contractors must be optimists. They have limited time to bid projects and are competing with their fellow contractors in a tight market. The courts are sympathetic to this problem. If during the prosecution of a claim the contractor can present documentation showing how it put its bid together and show the assumptions made as to means, methods and sequence, that documentation will have great weight as being the contemporaneous actions taken during a short bidding period based upon information available.

The basis for bid must not only be clear as to design intent, but also be a clear basis for constructability and biddability. Resident engineers and inspectors can provide a great service at this point by addressing the following:

- Specifications and divisions are appropriate and per a standard format.
- Procedures for substitutions are clear.
- Appropriate material and equipment standards are specified.
- No sole source or brand name materials or equipment are specified.
- Technologies and notations are consistent.
- Plans and specifications allow a broad selection of appropriate construction means, methods and techniques.
- Cross-reference of drawings and specifications are complete.
- Complete description is provided of anything to be furnished by the owner with a schedule of delivery.
- Definition of items of work to be provided by each contractor for multiple contracts is provided.
- Definition of quality control responsibilities of contractor and owner is provided, with clear statement of tests and access required.
- Submittal requirements are clear.
- Review period for submittals is identified and appropriate.
- The construction schedule is feasible and clearly defined with schedule interface points identified.
- Completion times are specified.
- Supplemental data is referenced.
- Disposal requirements of excess material waste cleanup are identified.
- Divisions of work are clearly identified at contractor interfaces.
- Drawings are sufficiently detailed and work is clearly defined.
- Structure of bid form, bid schedule, etc., are clearly defined and unambiguous.
- Proper units are used for bid items.

- Bid items are clear as to the scope that they cover.
- Bid quantities are reasonable for work scope defined.
- Bid items are coordinated with drawings and specifications.
- Measurement and payment mechanisms are clearly defined and reasonably coordinated with bid items.
- Change order procedure is spelled out and basis for adjustment is identified.

6.5 Management of Outside Design Professionals

Studies have shown that over 60 percent of claims have their basis in defective and deficient contract documents. Owners often use the engineering services of engineering and construction companies and private design firms for preparation of these contract documents. Such professionals are not deemed to be perfect and do not warrant that their plans and specifications are 100 percent free from errors and omissions. Liability of design professionals for errors and omissions is beyond the scope of this book.

Common causes of problems that owners experience with professional contracts include the following:

- Incomplete scope of work;
- Misunderstanding of work plan and the responsibilities of the parties;
- Unclear performance criteria;
- Interference and change by the owner;
- No internal quality assurance system by professional designer;
- Lack of independent reviews of professional work products;
- Lack of coordination among subconsultants;
- Inadequate selection procedure;
- Lack of an agreed-upon schedule for professional performance; and
- Conflict between the professional as an agent and the professional as an independent contractor producing work products.

Of all causes, the most common problem is a conflict of expectations between the parties. Professionals are contractors and must be managed by contract. This management involves development of a scope of work, performance criteria, budgets, and schedules—all of which form the basis for a meeting of the minds.

Much of the responsibility for the details of the scope of work should rest with the professional. Has the owner required that the professional submit a detailed work plan describing the who, what, where, when and how of the professional's approach to meet the requirements of the owner? Is there an adequate procedure for professional selection that identifies the professional firm that is best qualified to perform the scope of work?

A selection process should contain as a minimum the following:

- Internal review of the scope of work to determine that it meets the requirements of all parties within the owner's organization and adequately describes expectations.
- Requirements for a proposal format that includes: the approach to be taken, a work plan, personnel to work on the job, previous experience in similar work, references, and cost data as required by the owner.
- A formal review and selection process of an unbiased committee to screen written proposals and select three to five firms for oral interviews.
- Requirements that personnel who will work on the project present the orals rather than salesmen or marketing professionals.
- An independent committee that receives the oral presentations and makes recommendations to management for award.
- If procedures allow, selection should be made upon the basis of qualifications, personnel, technical competence, experience in similar work, etc. Negotiations can then be conducted with the top-rated firm to arrive at a price. If negotiations on price are not successful, then negotiations can move to the next rated firm on the best-qualified list. In any event, once selection is made, the contract must be managed by the owner. Changes must be controlled and work products subject to scrutiny for independent review, preferably by those who will be responsible for its construction in the field, *i.e.*, the resident engineer and the inspectors.

The role of the professional during construction should be specified. That role is usually limited to review of submittals, responses to requests for information and design clarifications, and occasional monitoring of work to determine that specific elements are being furnished and installed in accordance with the designer's intent. Owners must be careful when placing the design professional in a position as an agent of the owner, making decisions concerning the professional's work products. This occurs when the design professional is given construction management responsibilities. The concerns here are the inherent liability for construction problems resulting from design issues and the objectivity of the design firm in dealing with such problems.

Chapter 7

Avoiding Project Failure by Defining Requirements

Chapter 7

Avoiding Project Failure by Defining Requirements

Project owners often think of project initiation as the time when they have determined the project mandate and objectives and started to assemble a project team who will begin the work. Long before assembling the project team, however, project owners must assess the "Requirements" and value of the project or risk succumbing to the most common root cause of project failure. Before the project team is assembled, it is essential that project owners understand the business value that is being sought and what alternatives are available to deliver this value. If a project is a leading opportunity, then exactly what does the project need to deliver to produce the business value? This should then be expressed in terms of business and project goals that are specific, measurable, attainable, realistic, and time-bound. Owners can then turn these goals into Requirements that the project can use as input.

The Business Planning stage process is reasonably well understood. It involves identifying and understanding an opportunity, attaining alignment with stakeholders, considering alternate solutions and their various attributes, shortlisting the alternatives, and finally selecting one with which to proceed. Owners often believe that the immediate next step is to commission a project development team and to assemble a comprehensive plan that will deliver the unique product, service, or result that is needed to support the business plan. Owners often omit, however, consideration as to just what they are going to tell the project team that they want. If owners think in business value terms early in the process, then one of the deliverables to the project development team at the end of this Business Planning stage is a set of clear expectations and instructions that form a comprehensive set of "Requirements."

Too often, the Requirements articulation is left to the project management team, who have not been involved in the conceptualization of the business value to be delivered or the development of the conditions and capabilities that are required to be present in the final delivered product, service, or result.

This section is about those things that should be done early in business planning and project formulation to ensure that key stakeholders are completely aligned on the Requirements that, if satisfied by a project management team, will produce the business value envisioned by the solution selected in the business plan. The completion of Requirements development is an input to the Project Charter which, when approved, signals the end of the Project Initiation Phase and the start of the Planning Phase.

Examination of these topics brings the project sponsors to a clear and articulated understanding of the business value that will be delivered by a well-understood and defined project. In doing this, project owners seek to avoid the most common root cause of project failure, which is not thinking early enough in the process about project Requirements.

The following topics are discussed in this section:

- The Problem-Project Failure
- Project Requirements
- Project Life Cycle
- Who Should Participate?
- The Requirements Development Plan
- Summary

7.1 The Problem-Project Failure

Projects fail at an alarming rate. Across industries, the studies are virtually unanimous that somewhere between one quarter and three quarters of all projects fail. The range is somewhat dependent upon the definition of failure and whether they are considering the delivery of business value or simply the achievement of objectives stated in project terms. These high failure rates are reported worldwide across industries and represent enormous amounts of wasted money.

In 2014, the Project Management Institute (PMI) reported on research that determined that poor project initiation is documented as one of the leading causes of project failure.[25] For many organizations, effective business analysis is not an integral part of their project work. As a result, projects are not delivering the intended business value. Fully one third of projects did not meet their original goals and business intent. Root causes were identified as:

- Inaccurate Requirements gathering; and
- Poor Requirements management practices.

These root causes were second only to changing organizational priorities as a root cause of project failure.

---

[25] Project Management Institute, *PMI's Pulse of the Profession: The High Cost of Low Performance*, 2014. www.pmi.org/-/media/pmi/documents/public/pdf/learning/thought-leadership/pulse/pulse-of-the-profession-2014.pdf.

Independent Project Analysis (IPA) finds that four of every five oil and gas megaprojects are characterized as failures.[26] Studies of other sectors including minerals, chemicals, and power found that the failure rate was in the region of 50 percent. Projects that are successful are those that did a very thorough job in the front end. The root cause of these project failures is the omission of clearly defined and articulated Requirements for the project's unique product, service, or result that are aligned and driven by the delivery of business value.

Business value is a concept that is unique to each organization and includes tangible and intangible elements. Business value can be measured in short-term or long-term characteristics. It can include hard assets such as money or equipment, or softer assets such as goodwill, brand recognition, or public benefit. Whether an organization is a publicly traded company, a government agency, or a nonprofit, all enterprises conduct operations focused on achieving their goals and improving their organizational value.

Projects are used to achieve business value through creating a unique product, service, or result that supports or enables corporate objectives and organizational goals required by an organization's strategic plan.[27] Failure to achieve a requirement in terms of scope, schedule, or cost is one thing, but failure to deliver business value defeats the purpose of the project and the organization's strategic plan. If such failure happens often enough, it may threaten the viability of the business or the ongoing existence of the organization.

In formulating a project, owners are of course spending their efforts and resources on development of an opportunity, markets, relationships, and the underpinning economics. However, owners often do not spend enough time and resources on clearly understanding the business problem and ensuring that they have looked at a full range of potential solutions. Furthermore, if a project is the best solution, owners too often do not spend enough time on clearly articulating the project Requirements. It is the lack of identification and management of project Requirements that PMI and IPA have identified as a leading cause of project failure.

7.2 Project Requirements

What are project Requirements? A project Requirement is a condition or capability that is essential to be present in a product or service, or necessary to satisfy the business objectives and organizational goals, and the Requirement must be consistent with the

---

[26] E. Merrow; T. Haïdar, "4 Out of Every 5 Oil & Gas Megaprojects Fail. But Why?" The Boardroom, 2013. Oil and Gas Today. http://www.ipaglobal.com/podcast-megaprojects-what-can-oil-gas-learn-from-other-sectors.

[27] Project Management Institute, "Projects and Strategic Planning," *A Guide to the Project Management Body of Knowledge (PMBOK® Guide)*, 5th ed., Pennsylvania: Project Management Institute, Inc., 2013, p. 10.

strategic plan and intended business value. Requirements may address various critical needs including:

- Business Requirements
- Stakeholder Requirements
- Solution Requirements
- Functional Requirements
- Nonfunctional Requirements
- Transition Requirements
- Project Requirements
- Quality Requirements

The list can be expanded, but these are basic criteria as defined by PMI in various publications. Clear Requirements are essential to develop and align the project objectives with the strategy of the larger organization.

Is the problem that we do not know how to do this? Not at all! Most project sponsors can take a strategic plan with organizational goals and business objectives and then, in the context of a specific opportunity, compose them into specific Requirements, work statements, objectives, etcetera. The problem is that it takes time and resources to do this, particularly in a multi-stakeholder environment. Business development leaders do not plan for the time and resources, so the critical preliminary work is deferred to later stages in the development. Have you ever heard "It's too early for that"? Rather than thinking early in project requirement terms, the thinking is deferred.

We know generally what to do; the problem is we just do not do it. We have a feeling of being schedule driven and that we must get on with it! We often feel that we need to move past the business formulation part quickly and get on with the project. We think they (the project management team) can figure it out as they go. Exacerbating the problem, project managers have become accustomed to receiving poorly articulated instructions, so it has become routine to start without a clear understanding of what Requirements need to be met. We lack the discipline to make sure we know where we are going before we kick off the project management team. Ready; fire; aim.

To come up with good baseline Requirements, we need to build Requirements development and refinement into our business planning processes so that a well-articulated set of Requirements is a necessary outcome of business planning. If we can have the Requirements fully evaluated, and even approved by stakeholders before we kick off the project management team, we will have the project management team focused on the right outcomes and at the same time minimize a significant source of disruptive change and cost escalation.

Poorly defined Requirements lead to incorrect ideas about what the business planners had in mind, which means the product, service, or result may or may not solve the business problem. They lead to changing understandings on the part of the project management team, which result in wasted effort, lost time, escalation of costs, and extension of schedules.

There is a well-known relationship in which the ability to influence a project decreases and the cost of changes increases as the project moves forward. It is shown in Figure 7-1, which suggests that early in the process there is greater flexibility, and the cost of changes is low. However, flexibility is lost as options are defined and chosen, and the impact of making a similar change later escalates very quickly.

Figure 7-1: Escalating Impact of Changes as the Project Moves Forward

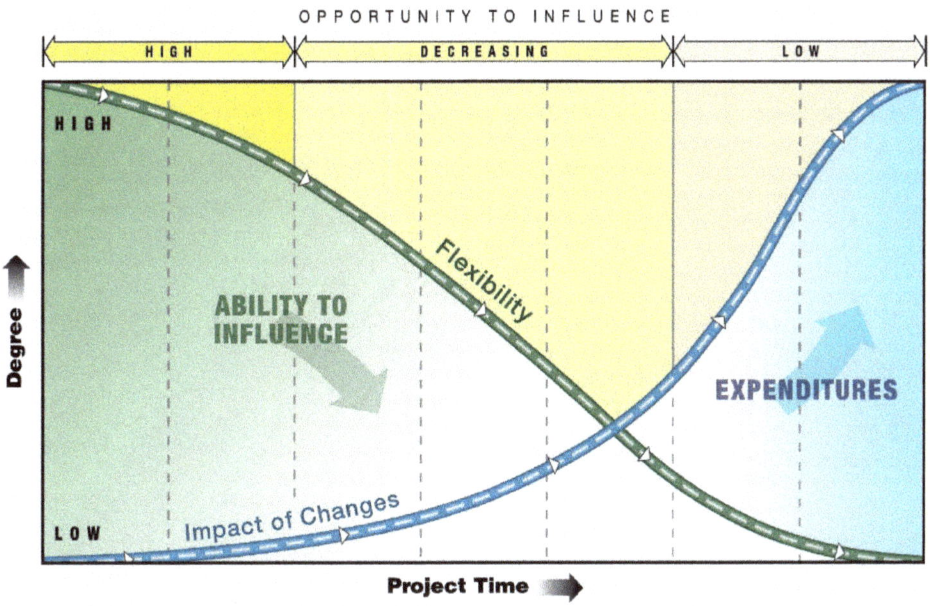

Successful projects have clear, well-defined project and business objectives. There is also a well-documented relationship between the clarity of objectives and overall project success. Projects that are only "somewhat clearly" defined rarely succeed, and even "fairly clearly" defined projects fail at a rate of two to one.[28]

As a result, we need a methodology that will guide us to think early enough in the process about the project's business value and lead us to develop well-articulated project Requirements.

[28] P. Luan, "How to Avoid Project Trainwrecks," *Oil and Gas Facilities Magazine*, Vol. 5, No. 2, pp. 25–28, April 2016.

7.3 Project Life Cycle

A project has a "Life Cycle" that can be defined in a series of phases it passes through from initiation to closure. As illustrated in Figure 7-2, all sequentially planned projects tend to have at least the following four phases:

1. Starting the project (often called Initiation)
2. Organizing and preparing (often referred to as Front End Loading)
3. Carrying out the project work
4. Closing the project

Figure 7-2: Discrete Stages of a Typical Project

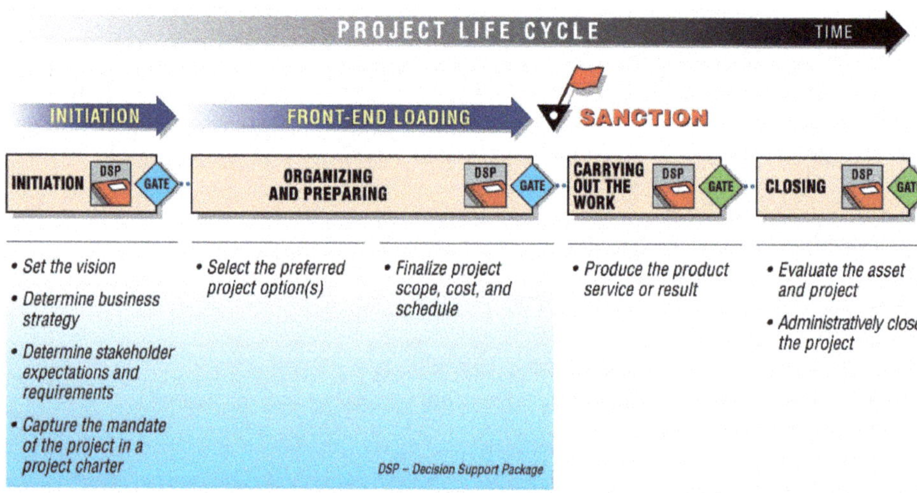

The purpose of organizing the project into phases is so that development proceeds in an orderly fashion. During Initiation phase, the definition of the project is much more conceptual than in later phases. The approach of developing more details as the project progresses from phase to phase is called progressive elaboration.

The Initiation phase begins early in the life of the business planning/entrepreneurial stage of the project development. Its purpose is to:

- Set the vision for the project.
- Define the initial scope.
- Commit initial financial resources.
- Identify internal and external stakeholders who will influence the overall outcome.

- Align the project's purpose with stakeholder expectations.
- Officially authorize the project through the approval of a Project Charter.

The Initiation phase ends with the approval of the Project Charter, which provides clear direction from stakeholders to the project management team about what is required and provides the approved basis upon which to begin work on the project.

There are a few key deliverables produced by the Initiation phase. These include:

- Business Case
- Project Statement of Work
- Project Charter

The Business Case is the document that explains *why* we should undertake these initiatives and any project that is part of the solution. It contains the argument for the action recommended. It is typically a succinct document. The intended audience is the organization's leaders, not the project management team. Nominally, the Business Case should include at least the following:

- A problem or opportunity statement that specifies the need for action.
- An analysis of the situation that reviews and demonstrates how the potential solutions support and contribute to the achievement of the organization's goals and objectives.
- A recommendation that presents the results of the feasibility analysis, constraints, assumptions, risks, and dependencies for each option, as well as the rationale for choosing a single option.[29]
- An evaluation that provides the plan for measuring the benefits realized at the conclusion of the project.

The project Statement of Work is a narrative description of products, services, or results to be delivered by the project. This document can capture product-oriented Requirements at a high level.

The Project Charter is a document that describes *what* the project is to deliver. It is an alignment document that authorizes the project to proceed on a specific mandate. The intended audience is stakeholders and the project team.

[29] Early in the project initiation process, it may be preferable to go forward with more than one option and a risk-based plan to select a single option at a specific milestone.

Nominally the Project Charter should include at least the following:

- Business need to be satisfied
- Assumptions
- Constraints
- Customer needs and high-level Requirements
- A summary of the Statement of Work defining the product, service, or result that the project is intended to deliver and related Requirements

The business development team should lead the preparation of these three documents. The project manager should participate as a functional member of that team.

We are looking for out of our Initiation phase is good-quality Requirements that address, at a high level, a full suite of stakeholder needs. The Requirements must, as far as practical, be stated in tangible terms so that they can be implemented, tracked, measured, and achieved. At the beginning, it is necessary to define all Requirements at a high level so that they can be composed in a hierarchical fashion into more specific Requirements as progressive elaboration of the project unfolds. If a new high-level Requirement is created later in the project, addressing it properly demands that the project reassess and ideally recycle back to the beginning. To avoid the chaos caused by change, and to avoid the cost escalation as well as schedule extension created by addressing change, it is essential to completely understand all Requirements as the project is being initiated.

Understanding the Requirements means being able to articulate them succinctly and being able to document a description of how these Requirements support the business need for the project. Once again, this starts out at a fairly high level in the Initiation phase. This forms the starting point for progressive elaboration and further breakout as details of the Requirements become known. Before commencing to carry out the project work, a final detailed baseline of the Requirements is needed. This baseline must be unambiguous (measurable and testable), traceable, complete, consistent, and acceptable to stakeholders. We must begin with this end in mind.

Thinking early enough in project Requirement terms means ensuring that key deliverables are identified in the Project Charter and Statement of Work. Organizations are often good at understanding business analysis and the selection of alternatives. They are often less adept at comprehensively documenting their Requirements, resulting in insufficient Requirements gathering.

7.4 Who Should Participate?

Often, the Business Case is prepared by a business development team, which does not include a representative who will be accountable for the project (typically the project manager). The completed Business Case, after approval, is presented to the project manager with a "do this" instruction. A Business Case is a document for executive review and is unlikely to contain the many details and Requirements that came up during the development. This information, in particular the Requirements, is lost to the project manager, if it was ever specifically developed in the first place. Thinking early enough in the process about project Requirements means developing an approach aimed at preventing this outcome.

If possible, the future project manager should be a participant in the business development team. The project manager is an invaluable resource when initiating the project to deliver information on project risks and cost/schedule alternatives. Also, the project manager is the future steward and leader of the initiative. The business development team should make sure that the project manager will accept accountability for the risks and cost/schedule requirements. Most importantly, however, the project manager will have an interest in driving the team towards articulation of clear Requirements.

The steps required are to:

- Clearly assign accountability for development of Requirements to the future project manager.
- Work towards a culture in which senior leaders' expectations include the creation of clear Requirements by the business development team as a valued outcome.
- Develop an approach whose objective is to produce well-thought-out, complete, and succinct Requirements. The plan supporting this approach would necessarily include allocation of time and resources to focus on Requirements development.

7.5 The Requirements Development Plan

We will use the "requirements elicitation" approach, described by PMI in its publication *Business Analysis for Practitioners: A Practice Guide*,[30] as a framework for describing what should be done. At the same time, this is offered as an example, and we should acknowledge that there are many ways to prepare and implement an approach that ensures we are thinking early enough in project terms to understand the Requirements of stakeholders.

[30] Project Management Institute, *Business Analysis for Practitioners: A Practice Guide*, Pennsylvania: Project Management Institute, Inc., 2015.

A central element to our approach is to decide whom we will consult and engage to determine Requirements. It is the stakeholder that we want to consult. "Stakeholders" refers to a diverse group of people not simply the organization's Sponsor who asked for this opportunity to be investigated. Stakeholders are:

> *An individual, group, or organization who may affect, be affected by, or perceive itself to be affected by a decision, activity, or outcome of the project.*[31]

This working definition of a stakeholder is used throughout the project in various contexts and necessarily encompasses a large group of people. When evaluating and developing Requirements this group is too large, and most do not have influence or interest at this stage. Therefore, we want to identify a subgroup that we will call "key stakeholders," whom we will consult for Requirements and alignment.

All key stakeholders may not have approval authority, but it is essential to recognize that they may have influence. It is essential to bear in mind that the Requirements belong to the stakeholders. It is their business value, they are the judges, and they determine success or failure. A project is simply a means to an end. The project manager's job is to satisfy stakeholders and deliver the product, service, or result that supports or enables business value. If these people's Requirements are not captured at this early stage, they have the potential to surface as required changes or developments later in the project with the attendant negative impacts. Including these Requirements up front is also central to a project manager's objective of minimizing change.

The key stakeholders are likely to be found among the following groups:[32]

- The project sponsor who is initiating and responsible for the project;
- Stakeholders who will benefit from an improved program or project;
- Stakeholders who will articulate and support the financial or other benefits of a solution;
- Stakeholders who will use the solution;
- Stakeholders whose role and/or activities performed may change as a result of the solution;
- Stakeholders who may regulate or otherwise constrain part or all of a potential solution;

---

[31] Project Management Institute, "Projects and Strategic Planning," *A Guide to the Project Management Body of Knowledge (PMBOK® Guide)*, 5th Edition, Pennsylvania: Project Management Institute, Inc., 2013, p. 563.

[32] Project Management Institute, *Business Analysis for Practitioners: A Practice Guide*, Pennsylvania: Project Management Institute, Inc., 2015.

- Stakeholders who will implement the solution; and
- Stakeholders who will support the solution.

Key stakeholders who are external to the organization may be called upon at this time for necessary consultation. Informal consultations may be held if including the stakeholders as part of the review group would be inappropriate. However, if they genuinely are key stakeholders, excluding them from the requirements development should be recognized as a risk for which a mitigation strategy may be needed.

Having identified which parties should be consulted, the next step is to be clear on why they will be consulted. The key stakeholders are consulted to develop a recommended path forward, which is documented in a business plan that is supported by a preliminary Statement of Work based on Requirements.

Recalling the introduction to this chapter, we related PMI's findings that the root causes of project failure are:

- Inaccurate Requirements gathering
- Poor Requirements management practices

Our objective of thinking early in the process in project business value terms means to recognize this risk and to respond to it by improving the Requirements gathering. This section focuses on the Requirements gathering and leaves the Requirements management practices during the execution of the project for a separate discussion.

The essence of Requirements gathering is to determine how and when to engage Stakeholders in elicitation activities as well as the methodology that will be utilized. The plan for Requirements elicitation should contain the following:

- The type of information that is to be elicited (Overall business? Operational? Technical or Functional?).
- Techniques that will be used to elicit the information.
- An identified sequence if there is a sequence (*e.g.*, general survey first, then workshop).
- A high-level plan for each session, including:
    - Specific objectives of each technique or session
    - Participants by Role or Name
    - Participant instruction, types of questions, or other material to focus people on the information being sought
- The type and timing of follow-up with participants.

There are numerous facilitation techniques that can be used and numerous references on how they can be used. Some examples include:

- Brainstorming
- Document analysis
- Facilitated workshops
- Focus groups
- Interviews
- Observation
- Prototyping
- Questionnaires and surveys

Leaders should remain alert to problems that could arise that might reduce the effectiveness of eliciting requirements. Each problem may require creativity and emotional intelligence to resolve. The types of issues might include:

- Conflicting perspectives and business needs from different stakeholders;
- Assumptions about what the solution should be, without first understanding the problem;
- Resistance to change, Requirement elicitation may require change management;
- Inability to step back from preconceived notions and solutions or the desire to press fast-forward because the answer is assumed to be obvious;
- Lack of access to the right stakeholders, either genuinely because of constraints or artificially because they decline to become engaged; and
- Lack of knowledge or experience on the part of stakeholders.

The elicitation is an iterative enquiry. This work should be done at the beginning of the project when the business analysis and plan are being assembled. There will be opportunities to elaborate further or progressively as the project gets underway. It is also important to consider when to stop further elicitation and how to identify when all the Requirements that drive business value have been identified.

Indicators of when to stop may include:

- The results are approved.
- An information model that guided the elicitation has been completed.
- The information being provided is redundant at the level being worked.

- There are no new top priority Requirements.
- How the defined business value is achieved is firmly supported by unambiguous Requirements.

It is essential to document the output of the elicitation activities. The documentation serves as a reference for what was requested by stakeholders and what they think is needed to accomplish the business purpose or maximize business value. It is, therefore, a primary reference and input for the project management team in defining what will and will not work, and what is and is not appropriate. It gives the project manager the basis to move forward with a minimum of disruptive changes that all cost time and money. It gives the project manager a baseline with which to maintain a stakeholder engagement plan in which Requirements are progressively elaborated throughout development, ending in verification that the final product, service, or result provides the stakeholders' Requirements.

The documentation of Requirements covers a broad array of issues and knowledge types, including the following.

- **Business Requirements** tend to be the goals, objectives, and higher-level needs of the organization. These might include economic, social, and environmental objectives.
- **Stakeholder Requirements** document the needs that each stakeholder might have. The "must haves," the "could haves," the "can't haves," and the "we don't wants."
- **Solution Requirements** are the functional and nonfunctional features, functions, and characteristics of the product, service, or result. This might include production capacities, reliability requirements, product specifications, user interfaces, maintainability, and many other parameters.
- **Transition Requirements** define how the organization will move from today's state to a future state in which the product, service, or result is in use. These requirements may address training, human resources, organizational management of change, or the way an organization switches over to the use of a new facility.
- **Project Requirements** describe how the project will be conducted, including:
    - How accurate do schedule and cost estimates need to be before approval;
    - Requirements in terms of compliance to specifications or quality standards (ISO9000);
    - Organization expectations regarding safety and environmental performance of the project; and

- Requirements that implement social or regulatory policy such as hiring practices, purchasing practices (*e.g.*, local content), or working with landowners.

- **Quality Requirements** are the conditions or criteria by which successful completion of a project deliverable will be validated. Examples are the possession of a permit issued by a regulator or the performance speed test record of a cargo ship.

At times, guidelines are helpful in assessing a Requirement. The following considerations are offered as a first-pass checklist. Typically, a Requirement should be:

- Written as a complete sentence (or sentences) including a subject, active verb, and object, and stating any conditions;
- Clear and unambiguous;
- Precise and concise. Each written Requirement only contains one Requirement, not a series of nested Requirements. If it can be broken into two, it should be;
- Consistent with other Requirements without contradiction or redundancy;
- Correct as confirmed by the relevant stakeholder;
- Measurable so that its attainment can be quantified and objectively validated or tested; and
- Traceable so that it can be used in Requirements tracking and management as well as change management. Any Requirements that are associated with higher level or over-arching Requirements should be documented as such. The Stakeholder(s) who is the source of the Requirement is identified so that the genesis of the Requirement can be traced.

7.6 Summary

Not thinking early enough in a project life cycle about project Requirements is a root cause of project failure. The research on this by leading authorities in project management is conclusive.

Project failure in this context is not failure of the project to meet its own objectives but rather the failure of the business to realize the business value or business objectives intended through undertaking a project.

7.6 – Summary

Projects are developed in phases, and the most important of these is the Initiation phase, in which the business can undertake the critical thinking that defines those things it wants a project to achieve. During the initiation phase, business planners should identify the conditions or capabilities (Requirements) that are essential for the product, service, or project result to deliver the intended business value.

Thinking early enough in the project life cycle about Requirements means including an objective in the initiation phase that high-level Requirements are gathered for later use by the project management team. A deliberate elicitation process is needed to gather the Requirements. With the Requirements in hand, project leadership is then ready to explain the mandate of the project to the project management team. Thinking early enough in the life cycle process about project Requirements helps to avoid a common root cause of project failure.

Chapter 8

Project Reviews

During the execution of a project, it is prudent for a contractor's senior management team to conduct periodic reviews of project performance to ensure that problems are being properly resolved, man-hour, cost and schedule information are being properly reported and analyzed to assure performance criteria are being met, changes are being properly estimated and sent to the owner for approval in a timely manner, comprehensive project documentation is being maintained, schedules are being updated and delays are being identified as to causes and responsibilities, and notices are being sent to the owner as required per the contract. In addition, major problems that should be brought to the attention of the owner at a higher level than the contractor's project manager should be dealt with immediately to mitigate or resolve significant issues.

Table 8-1 presents a detailed Contractor's Project Review Checklist. This checklist contains a lengthy list of questions that a senior-level project review team should use to ensure that all issues regarding project status are covered during project reviews.

Table 8-1: Contractor's Project Review Checklist (18 Pages)

LONG INTERNATIONAL

CONTRACTOR'S PROJECT REVIEW CHECKLIST

PROJECT NAME: ______________________________

____

PROJECT NO: ______________________________

DATE OF PROJECT REVIEW: ______________________________

PROJECT REVIEW TEAM SIGNATURES: ______________________________

____

____

____

____

____

____

____

____

____

____

____

Page 1 of 18

Table 8-1: Contractor's Project Review Checklist (continued)

LONG INTERNATIONAL

CONTRACTOR'S PROJECT REVIEW CHECKLIST

	PROBLEM / ISSUE	YES	NO	Unsure
	COST / FINANCIAL ISSUES			
☐	Are we invoicing to our Contracting Party in a timely manner?	☐	☐	☐
☐	Are our subcontractors invoicing us in a timely manner?	☐	☐	☐
☐	Are the cost accounts in sufficient detail to evaluate productivity?	☐	☐	☐
☐	Do we know our daily field overhead costs (staff, equipment, materials, expenses)?	☐	☐	☐
☐	Are our field overhead costs significantly over or under budget?	☐	☐	☐
☐	Are we tracking our field equipment rental costs?	☐	☐	☐
☐	Are we tracking our owned field equipment costs?	☐	☐	☐
☐	Are our field equipment costs within budget?	☐	☐	☐
☐	Are we tracking installed equipment costs by budget item?	☐	☐	☐
☐	Are our installed equipment costs within budget?	☐	☐	☐
☐	If we are over budget on installed equipment costs, do we know why?	☐	☐	☐
☐	Have we taken the appropriate measures to recover compensable extra installed equipment costs?	☐	☐	☐
☐	Are we tracking installed materials costs by budget item?	☐	☐	☐
☐	Are our installed materials costs within budget?	☐	☐	☐
☐	If we are over budget on installed materials costs, do we know why?	☐	☐	☐
☐	Have we taken the appropriate measures to recover compensable extra installed materials costs?	☐	☐	☐
☐	Do our costs accounts with which we are tracking labor correlate to specific schedule activities?	☐	☐	☐
☐	Is this correlation at the appropriate level of detail?	☐	☐	☐
☐	If a reasonable and sufficiently detailed correlation does not exist, can we create one?	☐	☐	☐
☐	Are our labor cost billings to date consistent with our internal reports for percent complete?	☐	☐	☐

Page 2 of 18

Table 8-1: Contractor's Project Review Checklist (continued)

LONG INTERNATIONAL

CONTRACTOR'S PROJECT REVIEW CHECKLIST

	PROBLEM / ISSUE	YES	NO	Unsure
COST / FINANCIAL ISSUES *(continued)*				
☐	Do we have a current cash flow projection?	☐	☐	☐
☐	Have cash flow projections taken any significant changes from earlier projections over the duration of the project?	☐	☐	☐
☐	Have we taken any measures to reduce retention?	☐	☐	☐
☐	Is there any contractual basis for reducing our retention?	☐	☐	☐
☐	Are we being paid promptly?	☐	☐	☐
☐	Have we taken any measures to expedite payment?	☐	☐	☐
☐	Are we paying our subcontractors per the contract terms?	☐	☐	☐
☐	Are we paying our vendors / suppliers per the contract terms?	☐	☐	☐
☐	Is a copy of the bid estimate in the project files?	☐	☐	☐
☐	Has the bid estimate been correlated to the control budget / cost account budget?	☐	☐	☐
☐	If so, does this correlation make sense?	☐	☐	☐

Page 3 of 18

Table 8-1: Contractor's Project Review Checklist (continued)

LONG INTERNATIONAL

CONTRACTOR'S PROJECT REVIEW CHECKLIST

PROBLEM / ISSUE		YES	NO	Unsure
LABOR / MAN-HOUR ISSUES				
☐	Are the man-hours on timesheets coded to specific cost accounts?	☐	☐	☐
☐	Are we tracking labor productivity by individual costs accounts?	☐	☐	☐
☐	Are we tracking progress by earned value?	☐	☐	☐
☐	Is our productivity acceptable?	☐	☐	☐
☐	Are we tracking the productivity of our subcontractors?	☐	☐	☐
☐	Is the productivity of our subcontractors acceptable?	☐	☐	☐
☐	Have we taken any measures to improve our productivity?	☐	☐	☐
☐	Have our productivity improvement measures been successful?	☐	☐	☐
☐	Is our Project Manager providing weekly labor and quantity budgets to the Foreman?	☐	☐	☐
☐	Do we have a baseline/as-planned manpower loading chart based on our original contract work?	☐	☐	☐
☐	Do we have a current manpower loading chart for the completion of the project?	☐	☐	☐
☐	Is our manpower level significantly higher or lower than the baseline manpower curve?	☐	☐	☐
☐	If it is, have we evaluated the productivity impact of such deviations?	☐	☐	☐
☐	Are we forecasting labor for future weeks by work activities?	☐	☐	☐
☐	Are we preparing three-week lookahead schedules for labor forecasts?	☐	☐	☐
☐	Are we preparing eight-week lookahead schedules for labor forecasts?	☐	☐	☐
☐	Have our labor forecasts deviated significantly from actuals?	☐	☐	☐
☐	Have we evaluated the productivity of future activities based on the productivity trends of like activities that already have some history?	☐	☐	☐
☐	Are copies of timesheets in the project files?	☐	☐	☐
☐	Are our timesheets signed by the Project Manager?	☐	☐	☐

Page 4 of 18

Table 8-1: Contractor's Project Review Checklist (continued)

LONG INTERNATIONAL

CONTRACTOR'S PROJECT REVIEW CHECKLIST

	PROBLEM / ISSUE	YES	NO	Unsure
	LABOR / MAN-HOUR ISSUES *(continued)*			
☐	Are copies of certified payrolls in the project files?	☐	☐	☐
☐	Are we working significant overtime?	☐	☐	☐
☐	Is our overtime consistent with our planned schedule and budget?	☐	☐	☐
☐	Are we being paid for our overtime premiums?	☐	☐	☐
☐	Are we being paid for any loss of productivity associated with overtime work?	☐	☐	☐
☐	Are we getting qualified craft from the labor market?	☐	☐	☐
☐	Are we getting adequate numbers of qualified craft from the labor market?	☐	☐	☐
☐	Are we experiencing high labor turnover?	☐	☐	☐
☐	Has our labor been attracted by other area projects with better incentives and working conditions?	☐	☐	☐
☐	Have we experienced a high rate of absenteeism?	☐	☐	☐

Page 5 of 18

Table 8-1: Contractor's Project Review Checklist (continued)

LONG INTERNATIONAL

CONTRACTOR'S PROJECT REVIEW CHECKLIST

PROBLEM / ISSUE		YES	NO	Unsure
QUANTITIES ISSUES				
☐	Are installed quantities being tracked and reported regularly?	☐	☐	☐
☐	Are the installed quantities being reported to the same cost accounts as are the man-hours?	☐	☐	☐
☐	Do our installed quantities reports correlate with quantities delivered to the field?	☐	☐	☐
☐	Are we maintaining an equipment / material receiving report?	☐	☐	☐
☐	Are we maintaining as-built drawings on a weekly basis?	☐	☐	☐
☐	Are the as-built drawings be checked against installed quantity data to ensure consistency?	☐	☐	☐

Page 6 of 18

Table 8-1: Contractor's Project Review Checklist (continued)

LONG INTERNATIONAL

CONTRACTOR'S PROJECT REVIEW CHECKLIST

PROBLEM / ISSUE		YES	NO	Unsure
FINANCIAL / COST REPORTING ISSUES				
Are the following types of financial / cost reports being regularly created and evaluated?				
☐	Cost of Labor Man-hours Report	☐	☐	☐
☐	Labor Productivity Report	☐	☐	☐
☐	Change Order Analysis Report	☐	☐	☐
☐	Purchase Order Status Report	☐	☐	☐
☐	Aged Payable Report	☐	☐	☐
☐	Cash Position Report	☐	☐	☐

Page 7 of 18

Table 8-1: Contractor's Project Review Checklist (continued)

LONG INTERNATIONAL

CONTRACTOR'S PROJECT REVIEW CHECKLIST

	PROBLEM / ISSUE	YES	NO	Unsure
	CHANGE ORDER ISSUES			
☐	Have we experienced a high number of Change Orders?	☐	☐	☐
☐	Was this level of Change Orders anticipated?	☐	☐	☐
☐	Are there an unusually high number of errors and omissions in the drawings and specifications?	☐	☐	☐
☐	Have the Change Proposals and Change Orders been approved in a timely manner?	☐	☐	☐
☐	Has there been a change in the contract work without the issuance of a formal Change Order?	☐	☐	☐
☐	Is the pricing of our Change Proposals and Change Orders current?	☐	☐	☐
☐	Are we pricing our Change Proposals and Change Orders per the terms of the contract?	☐	☐	☐
☐	Do our Change Proposals and Change Orders request time extensions?	☐	☐	☐
☐	Do we submit fragnets and schedule information with each Change Proposal or Change Order pricing?	☐	☐	☐
☐	Do our Change Proposals and Change Orders contain pricing for delay costs?	☐	☐	☐
☐	Do our Change Proposals and Change Orders contain pricing for impact / loss of productivity costs?	☐	☐	☐
☐	Does the Project have the commonly used reports on productivity impact factors for pricing Change Proposals and Change Orders?	☐	☐	☐
☐	Does our Project Manager understand productivity impact pricing?	☐	☐	☐
☐	Do our Change Proposals and Change Orders have adequate language to reserve our rights to additional costs?	☐	☐	☐
☐	Have we evaluated the reservation of rights language in Change Proposals and Change Orders from our subcontractors?	☐	☐	☐
☐	Does the subcontractor's reservation of rights language open us to claims?	☐	☐	☐
☐	Do our Change Orders from subcontractors provide full accord and satisfaction language?	☐	☐	☐

Page 8 of 18

Table 8-1: Contractor's Project Review Checklist (continued)

LONG INTERNATIONAL

CONTRACTOR'S PROJECT REVIEW CHECKLIST

	PROBLEM / ISSUE	YES	NO	Unsure
	CHANGE ORDER ISSUES *(continued)*			
☐	Have our Change Proposals and Change Orders been approved by our Contracting Party?	☐	☐	☐
☐	Are we invoicing for Change Order costs?	☐	☐	☐
☐	Are we tracking the man-hours and costs for Change Order Work separately?	☐	☐	☐
☐	Are we tracking the installed quantities for our Change Order work separately?	☐	☐	☐
☐	Have we approved Change Proposals and Change Orders submitted by our subcontractors?	☐	☐	☐
☐	Are our subcontractors invoicing for Change Order Work?	☐	☐	☐
☐	Have we performed any Extra Work / Work Tradeoffs for which there is no formal cost agreement?	☐	☐	☐
☐	Have we performed any Change Order Work without agreement and approval on costs?	☐	☐	☐

Page 9 of 18

Table 8-1: Contractor's Project Review Checklist (continued)

LONG INTERNATIONAL

CONTRACTOR'S PROJECT REVIEW CHECKLIST

	PROBLEM / ISSUE	YES	NO	Unsure
	OTHER ISSUES			
☐	Do we have personnel with the required contracting and professional licenses on the project or at the office responsible for the project?	☐	☐	☐
☐	Has work been directed in a manner or method that is different from or more expensive than the method that was originally anticipated?	☐	☐	☐
☐	Are we aware of any material facts that were withheld from our knowledge that would have affected our costs, time, or method of performance?	☐	☐	☐
☐	Was our work made impossible due to specification errors or unattainable performance requirements?	☐	☐	☐
☐	Has notice and a Change Order been provided to our contracting party for changed methods?	☐	☐	☐
☐	Have we been required to provide unanticipated engineering services?	☐	☐	☐
☐	Has notice and a Change Order been provided to our contracting party for these additional engineering services?	☐	☐	☐
☐	Has the supply of utilities for our work been adequate?	☐	☐	☐
☐	Is there a proactive Safety Program on this project?	☐	☐	☐
☐	Is safety receiving the required emphasis?	☐	☐	☐
☐	Has a safety review been performed on the project?	☐	☐	☐
☐	Are the number of lost time accidents and lost time work days above average?	☐	☐	☐
☐	Have we received any OSHA citations?	☐	☐	☐
☐	Are the OSHA citations in our files?	☐	☐	☐
☐	Has our or our subcontractor's safety record been acceptable?	☐	☐	☐
☐	Are we having regular meetings with our contracting party?	☐	☐	☐
☐	Are we having regular meetings with our subcontractors?	☐	☐	☐
☐	Is our scope of work well defined?	☐	☐	☐

Page 10 of 18

Table 8-1: Contractor's Project Review Checklist (continued)

LONG INTERNATIONAL

CONTRACTOR'S PROJECT REVIEW CHECKLIST

	PROBLEM / ISSUE	YES	NO	Unsure
	OTHER ISSUES *(continued)*			
☐	Do we have adequate (numbers and experience) field staff assigned to the project?	☐	☐	☐
☐	Have we experienced any work that was commercially impractical, i.e., unreasonable difficult in locating or obtaining necessary materials, supplies, tools, or equipment, or at an unreasonably high cost, or procurement or delivery was not possible at the time or place needed?	☐	☐	☐
☐	Have we unexpectedly experienced any work that was extremely expensive, dangerous or difficult?	☐	☐	☐
☐	Has any of our work or our subcontractor's work been damaged by others?	☐	☐	☐
☐	Have we provided formal written notice relative to this damaged work?	☐	☐	☐
☐	Has our contracting party representative been accessible on a day-to-day basis to address our problems and concerns?	☐	☐	☐
☐	Are our subcontractors bonded?	☐	☐	☐
☐	Do we have copies of their bonds?	☐	☐	☐
☐	Are the site housekeeping conditions adequate?	☐	☐	☐
☐	Does the project have on file an analysis of the required notices to protect our rights?	☐	☐	☐
☐	If required by law, have pre-lien notices been properly filed to the owner?	☐	☐	☐

Page 11 of 18

Table 8-1: Contractor's Project Review Checklist (continued)

LONG INTERNATIONAL

CONTRACTOR'S PROJECT REVIEW CHECKLIST

	PROBLEM / ISSUE	YES	NO	Unsure
	SCHEDULE ISSUES			
☐	Did the bid estimate contain a bid schedule?	☐	☐	☐
☐	Do we have a hard copy of this bid schedule?	☐	☐	☐
☐	Do we have a computer disk containing the files of this bid schedule?	☐	☐	☐
☐	Was the bid schedule man-loaded?	☐	☐	☐
☐	Was the bid schedule in CPM format?	☐	☐	☐
☐	Was the Notice-to-Proceed issued on the date set forth in the Contract?	☐	☐	☐
☐	Was a detailed as-planned schedule created after we started the project?	☐	☐	☐
☐	Is the detailed as-planned schedule consistent with the bid schedule?	☐	☐	☐
☐	Did we transmit a copy of our detailed as-planned schedule to our contracting party for approval?	☐	☐	☐
☐	Did we receive approval from our contracting party relative to our detailed as-planned schedule?	☐	☐	☐
☐	Did we adequately address all concerns relative to our as-planned schedule that were raised by our contracting party?	☐	☐	☐
☐	Did we plan on completing our work earlier than the contract completion date?	☐	☐	☐
☐	Did our as-planned schedule clearly show an early completion date?	☐	☐	☐
☐	Does the detailed as-planned schedule correlate with the chart of cost accounts at the appropriate level?	☐	☐	☐
☐	Is the any written documentation relative to the cost accounts / schedule activities' correlation in the project files?	☐	☐	☐
☐	Does the detailed as-planned schedule contain submittal and procurement activities?	☐	☐	☐
☐	Is there any evidence of acceleration when the bid schedule is compared to the detailed as-planned schedule?	☐	☐	☐
☐	Are we maintaining a statused as-built schedule?	☐	☐	☐

Page 12 of 18

Table 8-1: Contractor's Project Review Checklist (continued)

LONG INTERNATIONAL

CONTRACTOR'S PROJECT REVIEW CHECKLIST

	PROBLEM / ISSUE	YES	NO	Unsure
	SCHEDULE ISSUES *(continued)*			
☐	Is our as-built schedule current?	☐	☐	☐
☐	Is the statused as-built schedule consistent with the observed status in the field?	☐	☐	☐
☐	Do we have hard copies of all of our statused as-built schedules?	☐	☐	☐
☐	Do we have computer disks containing the files of all of our statused as-built schedules?	☐	☐	☐
☐	Is our Contracting Party maintaining a statused as-built schedule?	☐	☐	☐
☐	Is our Contracting Party's as-built schedule current?	☐	☐	☐
☐	Does the statused as-built schedule show the current contract completion and milestone dates including all current time extensions?	☐	☐	☐
☐	Do we have hard copies of all of our Contracting Party's statused as-built schedules?	☐	☐	☐
☐	Do we have computer disks containing the files of all of our Contacting Party's statused as-built schedules?	☐	☐	☐
☐	Are all known delays to our work included in the statused as-built schedule?	☐	☐	☐
☐	Is a Daily Schedule Activity log being prepared?	☐	☐	☐
☐	Is the duration and linkage of the delays to the affected activities accurate?	☐	☐	☐
☐	Are delays relative to late RFI responses in the statused as-built schedule?	☐	☐	☐
☐	Have we provided formal written notice to our contracting party for late RFI responses?	☐	☐	☐
☐	Are delays relative to late submittal responses in the statused as-built schedule?	☐	☐	☐
☐	Have we provided formal written notice to our contracting party for late submittal responses?	☐	☐	☐
☐	Are delays relative to Change Orders in the statused as-built schedule?	☐	☐	☐

Page 13 of 18

Table 8-1: Contractor's Project Review Checklist (continued)

LONG INTERNATIONAL

CONTRACTOR'S PROJECT REVIEW CHECKLIST

	PROBLEM / ISSUE	YES	NO	Unsure
	SCHEDULE ISSUES *(continued)*			
☐	Are delays relative to late delivery of equipment or materials by others in the statused as-built schedule?	☐	☐	☐
☐	Have we provided formal written notice to our contracting party for late equipment or material deliveries?	☐	☐	☐
☐	Have there been any differing site conditions, i.e., different from those described in the contract or that are unusual or extraordinary, that have delayed our work and / or have increased our costs?	☐	☐	☐
☐	Have we provided formal written notice to our contracting party for all differing site conditions problems?	☐	☐	☐
☐	Have there been significant delays to predecessor work by others?	☐	☐	☐
☐	Have we provided formal written notice to our contracting party for late predecessor activities?	☐	☐	☐
☐	Are we using an appropriate scheduling technique to evaluate the impact of delays to our work, such as time impact analysis?	☐	☐	☐
☐	Have we lost float as a result of the delays to our work?	☐	☐	☐
☐	Has the loss of float caused or will it cause stacking of trades?	☐	☐	☐
☐	Have we requested time extensions for all known delays?	☐	☐	☐
☐	Have we received any time extensions?	☐	☐	☐
☐	Are the time extensions consistent with our entitlement for delay?	☐	☐	☐
☐	If we have not received adequate time extensions, have we transmitted notice letters?	☐	☐	☐
☐	Have we incurred weather and / or force majeure delays?	☐	☐	☐
☐	Have we included these weather and / or force majeure delays in the statused as-built schedule?	☐	☐	☐
☐	Have these weather and / or force majeure delays increased our costs?	☐	☐	☐
☐	Are we being directed to perform our work earlier than the schedule would require?	☐	☐	☐

Page 14 of 18

Table 8-1: Contractor's Project Review Checklist (continued)

LONG INTERNATIONAL

CONTRACTOR'S PROJECT REVIEW CHECKLIST

	PROBLEM / ISSUE	YES	NO	Unsure
SCHEDULE ISSUES *(continued)*				
☐	If we are being directed to accelerate, have we written letters to our contracting party to request additional costs?	☐	☐	☐
☐	Are we being constructively accelerated, i.e., are justifiable time extensions being ignored?	☐	☐	☐
☐	Have we written notices relative to this constructive acceleration?	☐	☐	☐
☐	Are we tracking the progress of Change Order work separately in the schedule?	☐	☐	☐
☐	Were there any delays in the timely receipt of construction drawings prepared by others?	☐	☐	☐
☐	Did we provide notice for these delays in the receipt of construction drawings that were prepared by others?	☐	☐	☐
☐	Has our work space been congested by the storage of materials by others?	☐	☐	☐
☐	Have we provided formal written notice of this stored material congestion?	☐	☐	☐
☐	Have we experienced delays in securing permits or right-of-ways?	☐	☐	☐
☐	Have we provided formal written notice relative to these permit or right-of-way delays?	☐	☐	☐
☐	Have we experienced delays to the access of our work?	☐	☐	☐
☐	Have we provided formal written notice relative to these access delays?	☐	☐	☐
☐	Were there any strikes or work stoppages?	☐	☐	☐
☐	Has there been defectively installed work by others that has caused delay in the installation of our work?	☐	☐	☐
☐	Have we provided formal written notice relative to this defectively installed work by others?	☐	☐	☐
☐	Have we caused any delay to our work?	☐	☐	☐
☐	Has our subcontractors caused any delay to their work?	☐	☐	☐

Page 15 of 18

Table 8-1: Contractor's Project Review Checklist (continued)

LONG INTERNATIONAL

CONTRACTOR'S PROJECT REVIEW CHECKLIST

	PROBLEM / ISSUE	YES	NO	Unsure
SCHEDULE ISSUES *(continued)*				
☐	Have our subcontractor's delays been evaluated for overall project delays?	☐	☐	☐
☐	Have we experienced delay in the receipt of equipment and / or materials to the project?	☐	☐	☐
☐	Have these delays affected the critical path and could delay the project?	☐	☐	☐
☐	Have we experienced any delay or unreasonableness in the inspection of our work by others?	☐	☐	☐
☐	Have we provided formal written notice relative to these inspection delays?	☐	☐	☐

Page 16 of 18

Table 8-1: Contractor's Project Review Checklist (continued)

LONG INTERNATIONAL

CONTRACTOR'S PROJECT REVIEW CHECKLIST

PROBLEM / ISSUE	YES	NO	Unsure
DOCUMENTATION			
☐ Are the project files well organized and in good order?	☐	☐	☐
☐ Is current documentation being regularly filed?	☐	☐	☐
☐ Is the following documentation in the project files?			
☐ Copy of the Contract	☐	☐	☐
☐ Copy of the Subcontracts	☐	☐	☐
☐ Copy of the original estimate	☐	☐	☐
☐ Site investigation report / checklist	☐	☐	☐
☐ Daily construction reports	☐	☐	☐
☐ RFIs	☐	☐	☐
☐ RFI log	☐	☐	☐
☐ Submittals	☐	☐	☐
☐ Submittal log	☐	☐	☐
☐ Correspondence	☐	☐	☐
☐ Correspondence log	☐	☐	☐
☐ Change Orders	☐	☐	☐
☐ Change Proposals	☐	☐	☐
☐ Change Order log	☐	☐	☐
☐ Change Proposal log	☐	☐	☐
☐ Invoices to our contracting party	☐	☐	☐
☐ Invoices from our Subcontractors	☐	☐	☐
☐ Drawings	☐	☐	☐
☐ Specifications	☐	☐	☐
☐ Bid schedule	☐	☐	☐
☐ Detailed as-planned schedule	☐	☐	☐
☐ Schedule updates	☐	☐	☐
☐ Schedule data disks	☐	☐	☐
☐ Daily Schedule Activity log	☐	☐	☐
☐ Schedule meeting minutes	☐	☐	☐
☐ Other meeting minutes	☐	☐	☐

Page 17 of 18

Table 8-1: Contractor's Project Review Checklist (continued)

LONG INTERNATIONAL

CONTRACTOR'S PROJECT REVIEW CHECKLIST

PROBLEM / ISSUE		YES	NO	Unsure
DOCUMENTATION *(continued)*				
☐ Is the following documentation in the project files? *(continued)*				
	☐ Project controls reports	☐	☐	☐
	☐ Internal progress reports	☐	☐	☐
	☐ External progress reports	☐	☐	☐
	☐ Monthly cost / financial reports	☐	☐	☐
	☐ Monthly cash flow projections	☐	☐	☐
	☐ Drawing revision log	☐	☐	☐
	☐ Timesheets	☐	☐	☐
	☐ Certified payrolls	☐	☐	☐
	☐ Foreman's diaries	☐	☐	☐
	☐ Project Manager's diaries	☐	☐	☐
	☐ Purchase orders	☐	☐	☐
	☐ Issue files	☐	☐	☐
	☐ As-built drawings	☐	☐	☐
	☐ Backup tapes from field computers	☐	☐	☐
	☐ Field equipment lists with on-off dates	☐	☐	☐
	☐ Accident reports	☐	☐	☐
	☐ OSHA citations and logs	☐	☐	☐
	☐ Nonconformance reports	☐	☐	☐
	☐ Progress Photos, dated and noted	☐	☐	☐
	☐ Time and Materials work log	☐	☐	☐
	☐ Labor agreement	☐	☐	☐
	☐ Certificates of Insurance (ours and our subcontractor's)	☐	☐	☐
	☐ Bonds (ours, our subcontractor's)	☐	☐	☐
	☐ Equipment and tool inventories	☐	☐	☐
	☐ Equipment maintenance logs	☐	☐	☐
	☐ Start-up / Testing plan	☐	☐	☐
	☐ Material control receiving tickets	☐	☐	☐
	☐ Project close out plan	☐	☐	☐
	☐ QC program	☐	☐	☐

Page 18 of 18

Chapter 9

Site Investigation

A thorough site investigation by the contractor is essential to ensure all site-specific information is collected for preparation of the bid estimate. If the contractor fails to perform an adequate site investigation, many of the impacts that may affect the contractor's time and cost of performance may not be recognized. If such conditions would normally be identified and recognized by a prudent and experienced contractor through a site investigation, recovery from unanticipated costs caused by these impacts would normally be precluded.

A thorough site investigation would include the following categories of information:

- General site information, *i.e.*, soil conditions, utilities, subsurface conditions
- Detailed surface conditions
- Detailed subsurface conditions
- Permits, fees, and tax requirements
- Material staging and site logistics
- Labor information
- Weather data
- Equipment and materials information
- Transportation information
- Pricing data
- Notes of meetings with owner's site representatives

Table 9-1 shows an example of a Site Investigation Checklist.

Table 9-1: Site Investigation Checklist (9 Pages)

LONG INTERNATIONAL

SITE INVESTIGATION CHECKLIST

The following checklist can be effectively used during the estimating phase. In addition, *pictures, memoranda, and minutes of meetings* with the owner's representative(s) should be maintained in the bid file.

PROJECT NAME: __________

PROJECT LOCATION: __________

ELEVATION: __________ BAROMETRIC PRESSURE: __________

1. GENERAL:

Soil (*Brief description of stability, type, gradation, density*) __________

Identify Underground Utilities (*Check*)

- ☐ Utilities
- ☐ Power
- ☐ Water
- ☐ Telephone
- ☐ Gas

General Description of Subsurface Conditions (*Check*)

- ☐ Hidden Foundations
- ☐ Rock
- ☐ Blasting Required
- ☐ Ground Water Level/Elevation
- ☐ Hazardous Waste
- ☐ Excavation
- ☐ Other

Remarks __________

YES	NO	
☐	☐	Is further investigation needed?
☐	☐	Is piling required?
☐	☐	Has soil report been obtained from owner?
☐	☐	Have we performed our own site investigation?

Page 1 of 9

Table 9-1: Site Investigation Checklist (continued)

LONG INTERNATIONAL

SITE INVESTIGATION CHECKLIST

1. **GENERAL**: (*continued*)

YES NO
- ☐ ☐ Have we checked with local soils engineer?
- ☐ ☐ Have we taken pictures of site?
- ☐ ☐ Have aerial photographs been obtained?
- ☐ ☐ Have topographic maps been obtained?
- ☐ ☐ Have we made inquiries of owner regarding any questions we have?

2. **SURFACE CONDITIONS:**

Type and amount of clearing and demolition *(Describe problems with site clearing / grading)*

____

____

YES NO
- ☐ ☐ Is site accessible?
- ☐ ☐ Are all utilities available? Do we have to pay for?

  YES NO
  - ☐ ☐ Potable water
  - ☐ ☐ Non-potable water
  - ☐ ☐ Natural gas
  - ☐ ☐ Power
- ☐ ☐ Are other contractors working in area?
- ☐ ☐ Are there other interferences?

  ____

  ____
- ☐ ☐ Will topsoil have to be removed?
- ☐ ☐ Can excess soil be disposed of on-site?
- ☐ ☐ Will access roads require maintenance?
- ☐ ☐ Is there a stream, swamp, or lake in the immediate area?
- ☐ ☐ Is there a structure which will affect construction operations (such as a nearby building which could affect crane operations, etc.)?
- ☐ ☐ Is borrow area easily accessible?

Page 2 of 9

Table 9-1: Site Investigation Checklist (continued)

LONG INTERNATIONAL

SITE INVESTIGATION CHECKLIST

2. SURFACE CONDITIONS: *(continued)*

YES	NO	
☐	☐	Is the borrow quantity that is available adequate for the job?
☐	☐	Must borrow/excess soil be hauled from off-site?
☐	☐	How far must borrow be hauled?
☐	☐	Will road conditions to borrow area require special restrictions, maintenance, or equipment?
☐	☐	Are there royalty or other costs necessary for borrow materials?
☐	☐	Are there any special safety problems (high lines, etc.)?
☐	☐	Are there any noise abatement requirements or working hour restrictions?
☐	☐	Are elevations apparently as represented in drawings?
☐	☐	Are there any activities or operations by owner which can affect construction work?
☐	☐	Is adequate space available for equipment, material, storage, and field office?
☐	☐	Is observation of site consistent with conditions shown on plans and drawings?

3. SUBSURFACE CONDITIONS:

YES	NO	
☐	☐	Are there sufficient test holes shown in bid documents?
☐	☐	Is there sufficient data about subsurface conditions shown in bid documents?
☐	☐	Have we examined existing cuts and pits at the site?
☐	☐	Have we drilled our own test holes?
☐	☐	Have special measures been taken for protecting against flooding during construction?
☐	☐	Are there restrictions on pumping water?
☐	☐	Can we dispose of the water?
☐	☐	Have special measures been taken for blasting of rock, over-excavation, etc.?
☐	☐	Have special measures been taken for shoring of ditches?
☐	☐	Are underground utilities identified in bid documents?

Page 3 of 9

Table 9-1: Site Investigation Checklist (continued)

LONG INTERNATIONAL

SITE INVESTIGATION CHECKLIST

3. SUBSURFACE CONDITIONS: *(continued)*

YES NO
☐ ☐ Do the bid documents contain a rock clause?

☐ ☐ Do the bid documents contain a clause compensating you for flooding or other acts of God?

☐ ☐ Are there nearby rivers and streams? Are there hazards from high water?

☐ ☐ Do the bid documents contain other protections against contingencies?

☐ ☐ Did investigation disclose any inconsistencies with conditions shown in plans and drawings?

List site contingencies or risks that must be considered in bid price
____
____

4. PERMITS, FEES, AND TAXES: *(Check)*

List Permits and Fees required, and show cost:

☐ Building Permit Cost: ________________
☐ Electricity Cost: ________________
☐ Water Cost: ________________
☐ Gas Cost: ________________
☐ Telephone Cost: ________________
☐ Other Cost: ________________

YES NO
☐ ☐ Do the utility providers need considerable advance notice to provide service to the site?

Who will provide electric transformers? ____________________

Describe construction water source characteristics. ____________________
____
____

YES NO
☐ ☐ Is there a local drinking water hauling contractor?

☐ ☐ Is adequate water pressure available?

☐ ☐ Is the project tax exempt? If not tax exempt, indicate applicable taxes:
State Sales (Use) Tax ________%
Local Sales (Use) Tax ________%
Other Tax ________%

☐ ☐ Are there other special fees or licenses that will be required? If so, list.
____
____

Page 4 of 9

Table 9-1: Site Investigation Checklist (continued)

LONG INTERNATIONAL

SITE INVESTIGATION CHECKLIST

5. LABOR:

YES NO

☐ ☐ Is local, skilled labor readily available?

☐ ☐ Is housing for transient labor camp necessary?

☐ ☐ Will overtime shift work be required?

☐ ☐ Are there other major projects coming up in the area which could affect labor market?

☐ ☐ Are there similar projects under construction in the area?

☐ ☐ Do we have the wage rates? Fringe benefits?

☐ ☐ Are labor rates due to increase (new union agreements, etc.)?

☐ ☐ Are strikes possible due to expired labor contracts?

☐ ☐ Is the project being performed in an area or under conditions which could cause inefficiencies, i.e., poor labor market, availability of labor, etc.?

What productivity should be expected from labor in the area in relationship to our basic unit rates? ______________________

☐ ☐ Will travel time be required?

☐ ☐ Is busing required?

☐ ☐ Are there any special potential jurisdictional problems?

☐ ☐ Is the construction area congested?

How many sides open? ____

☐ ☐ Is access possible?

6. WEATHER:

YES NO

☐ ☐ Has local climatological data obtained?

Rainfall Data	Jan – Mar	_____ Inches/Month
	Apr – Jun	_____ Inches/Month
	Jul – Sep	_____ Inches/Month
	Oct – Dec	_____ Inches/Month

Maximum Storm Rainfall in 24 hrs _____ inches in ________ (Indicate Month)

Page 5 of 9

Table 9-1: Site Investigation Checklist (continued)

LONG INTERNATIONAL

SITE INVESTIGATION CHECKLIST

6. WEATHER: *(continued)*

Maximum Storm Rainfall in One Hour Period _____ inches in ________ (Indicate Month)

Average Wind Velocity __________ mph _____ days/year

Average Storm Wind __________ mph _____ days/year

Temperature Range (Minimum and Maximum Range):

Jan – Mar	_____ °F – _____ °F
Apr – Jun	_____ °F – _____ °F
Jul – Sep	_____ °F – _____ °F
Oct – Dec	_____ °F – _____ °F
Extreme Cold:	_____ °F Extreme Heat: _____ °F

Snow: Average inches per year_______ Maximum Storm_________inches
Snow Load (psf) _________________

Humidity		
	Jan – Mar	_____ %
	Apr – Jun	_____ %
	Jul – Sep	_____ %
	Oct – Dec	_____ %

Any: Sandstorms? YES ☐ NO ☐ Sleet? YES ☐ NO ☐ Heavy Fog? YES ☐ NO ☐ Tornadoes? YES ☐ NO ☐

Frost Depth ________

Electrical Storm Frequency ________

Earthquake Frequency ________ Richter Scale Severity________

Are there Dust, Mildew, or Pest Problems? YES ☐ NO ☐ Indicate ________________________

How much down time can we expect based on weather (e.g., 10-year climatological report from weather bureau)?________________

Will winter work be required? YES ☐ NO ☐

What special measures must be taken to protect against high winds?____________________

What special measures must be taken to protect against flooding?______________________

What special measures must be taken to protect against freezing?______________________

What special measures must be taken to protect against snow?_________________________

What special measures must be taken to protect against rain?__________________________

Page 6 of 9

Table 9-1: Site Investigation Checklist (continued)

LONG INTERNATIONAL

SITE INVESTIGATION CHECKLIST

7. **EQUIPMENT/MATERIALS**:

List special equipment that may be required for the project. ______

YES	NO	
☐	☐	Is nature of site (remote, difficult terrain) such that maintenance or depreciation factors must be adjusted?

List any special problem with transporting equipment to job site. ______

YES	NO	
☐	☐	Are there adequate local material suppliers?
☐	☐	Are there overhead obstructions that will interfere with construction equipment?
☐	☐	Are there any underground lines or open ditches that cannot be crossed with construction equipment?
☐	☐	Will we have sufficient area for storage of equipment and materials? List ______

Temporary Construction Facilities: (*Furnished by*)
- Office Trailer ______
- Warehouse ______
- Covered Storage Area ______
- Fenced Area ______

☐	☐	Are there fabrication shops in the local area? Describe capabilities.

8. **TRANSPORTATION**:

YES	NO	
☐	☐	Does location or schedule dictate special transportation (air freight, barge, etc.)?
		Where is nearest paved road? ______
☐	☐	Are there any railroad sidings near the project site? How far away? ____ mi.
☐	☐	Any pending transportation strikes? ______
☐	☐	What is local haulers capacity? ______
☐	☐	Is there special equipment available? ______
☐	☐	Is there a wharf at site? ______
☐	☐	Is dredging required for access? ______

Page 7 of 9

Table 9-1: Site Investigation Checklist (continued)

LONG INTERNATIONAL

SITE INVESTIGATION CHECKLIST

8. TRANSPORTATION: (*continued*)

YES NO
☐ ☐ Will site ambulance be required?

☐ ☐ Are site first aid facilities required?

Where is the nearest hospital? ____________________

List any special transportation problems ____________________

9. PRICES:

YES NO
☐ ☐ Are there any price hikes (steel, plywood, wire, pipe, etc.) pending?__________

☐ ☐ Are there any pending industry strikes that could escalate prices or cause material/equipment shortages and delays?__________

☐ ☐ Is an escalation clause included?

10. COMMUNICATIONS:

YES NO
☐ ☐ Are there telephone lines available?

☐ ☐ Is radio telephone needed and/or practical?

☐ ☐ Is a radio license required?

☐ ☐ Is mobile phone service available?

☐ ☐ Can we get a modem and/or fax line?

☐ ☐ Can we get high-speed internet access?

☐ ☐ Is there regular express mail service?

☐ ☐ Is it advantageous to open a post office box in the nearest town or city?

☐ ☐ Are overnight shipping services available in the nearest town or city?

Page 8 of 9

Table 9-1: Site Investigation Checklist (continued)

LONG INTERNATIONAL

SITE INVESTIGATION CHECKLIST

11. MEETINGS:

Summarize meetings, telephone conversations, and discussions with owner or its representative.

DATE	CONTRACTOR'S PERSONNEL	OWNER'S PERSONNEL	SUMMARY OF MEETING

Page 9 of 9

Chapter 10

Claims Mitigation During Construction

Successful claims avoidance results from prudent management activities. The following activities during the construction phase of a project are essential for both the owner's team and the contractor's team to mitigate claims and ensure the overall success of the project:

- Read and understand the contract documents.
- Implement a document control system to capture, code and file documents.
- Hold preconstruction meetings and reach agreements on key project objectives.
- Prioritize the relative importance of each objective.
- Define clearly the roles and responsibilities of each party.
- Allocate risks to the party best able to control those risks and provide equitable rewards for assuming risks.
- Develop performance criteria to communicate expectations and to measure each party's achievements.
- Coordinate activities involving several parties.
- Implement cost, schedule and quality control procedures.
- Hold periodic progress reviews and inspections.
- Maintain open communications throughout the project.

Commonly, a contractor's change order procedure may not require identification of delays to specific schedule activities and the completion date as part of the evaluation of the changed work. Schedule fragnets (a group of activities with logic) showing these time impacts to existing activities and the inclusion for new activities should be required as part of the change order procedure. In addition, the change estimate should specifically address the cost of potential delay and productivity loss cost impacts on other work as part of the change order estimate. If these estimates are not practical, the contractor should specifically reserve its rights to request additional compensation for delay and impact costs associated with change orders as part of the approval process.

Relative to a contractor's scheduling procedure, we suggest that, in addition to identifying the variances in durations and lag relationships when performing schedule updates for monthly reviews, the contractor's project personnel should also be required to identify the cause and responsibility for these delays as part of the schedule report for internal management review only. Causes of delay that are the owner's responsibility should be included in the contractor's progress reports to the owner.

Relative to cost control procedures, the contractor should track the man-hours and costs of changed work in the field. This is particularly important on lump-sum projects in which there may be disagreement as to the costs and impacts of changed work. Separate cost codes and tracking systems should be implemented to capture these increased costs. The contractor's procedure should also ensure that the level of detail in its base contract work is sufficient to capture any impact costs to base work caused by changed work. For example, if the contractor is only capturing the costs of piping in one cost code, and multiple changes start occurring to the piping systems, the contractor should consider expanding the piping cost codes to an area or system basis to ensure that cost impacts can be tracked.

All parties have obligations and duties under their contracts. Principal duties of the engineer acting as an agent of the owner are to make decisions and to interpret the contract when required. Under the principle of implied warranty, the engineer acting as an agent for the owner has a responsibility to cooperate affirmatively to keep the project moving. This cooperation often requires decisions in the following areas:

- Scope of work
- Clarifications
- Alleged errors and omissions
- Differing site conditions
- Submittals
- Or-equal requests
- Questions of quality and performance
- Schedule administration
- Requests for time extension
- Changes
- Questions of progress and early completion
- Third-party coordination
- Payment

On traditional construction projects involving an owner, design professional, and construction contractor, there exist overlapping responsibilities for decisions regarding the areas listed above, as shown in Figure 10-1.

Figure 10-1: Overlapping Roles of Major Parties in Traditional Construction Delivery

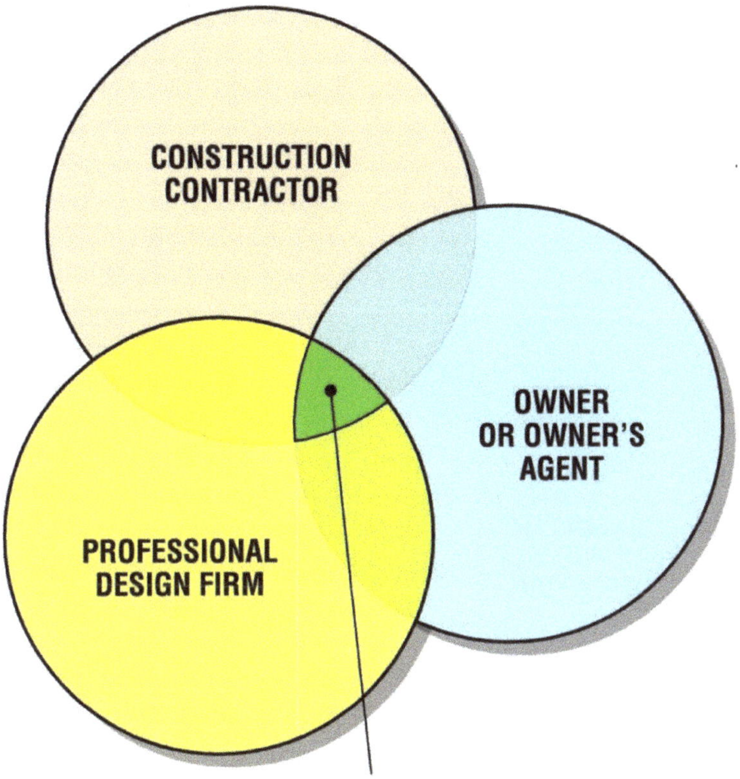

"Decisions" Regarding:

- Clarifications
- Errors & Omissions
- Scope of Work
- Differing Site Conditions
- Submittals and "Or Equals"
- Quality & Performance
- Time Extensions
- Progress & Early Completion
- Changes
- Third-Party Coordination
- Payment

On design-build projects, the design professional and construction contractor functions become the responsibility of one organization. As a result, design clarifications and the resolution of errors and omissions become problems that are resolved internally by the design-build contractor, as shown in Figure 10-2.

Figure 10-2: Overlapping Roles of Major Parties in Design-Build Approach

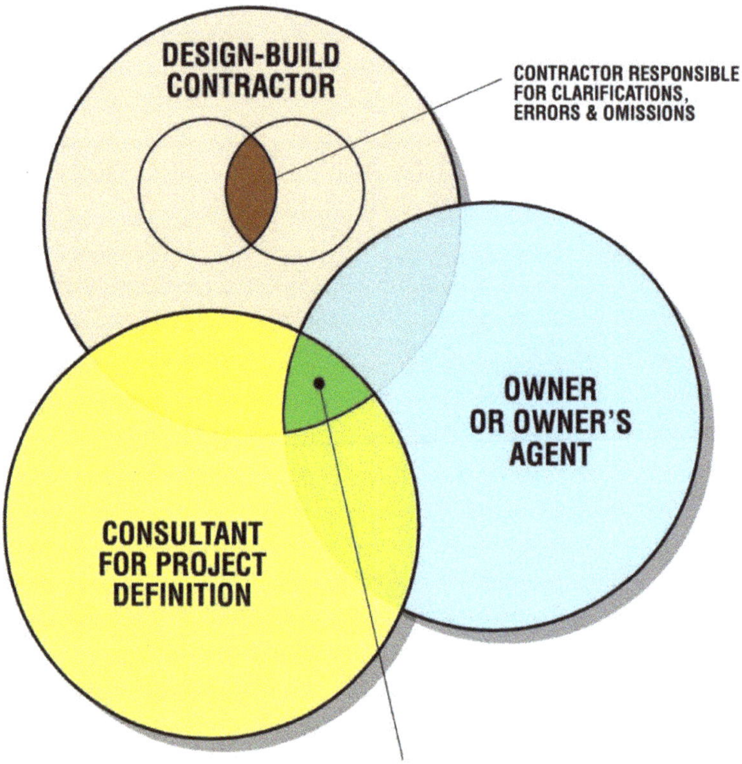

"Decisions" Regarding:

- Project Definition Changes
- Differing Site Conditions
- "Or Equals"
- Quality & Performance
- Time Extensions
- Progress & Early Completion
- Changes
- Third-Party Coordination
- Payment

Under either a traditional or design-build organizational structure, a decision management matrix should be implemented to clearly define who has review and/or approval responsibility for the various decision points during the project. Written procedures and assignment of responsibilities should be made for each area of overlapping responsibility. For example, the resident engineer should avoid the trap of becoming a historical scribe merely recording submittals as they are received. Under a decision management system, an individual should be designated to review the submittal file and to follow up on those submittals that have been in review longer than ten days.

Timely response is an essential element of a claims mitigation program. The courts look upon the engineer who takes timely and positive action with favor, as opposed to one who merely records data for historical purposes and stonewalls legitimate questions of interpretation or clarification.

Finally, a formal system of dispute resolution procedures should be developed, which encourages settlements. The goal of the parties should be to resolve disputes before they turn into claims leading to arbitration or litigation. Figure 10-3, "Claim Resolution Process," presents an example of a formal dispute resolution procedure.

Figure 10-3: Claim Resolution Process

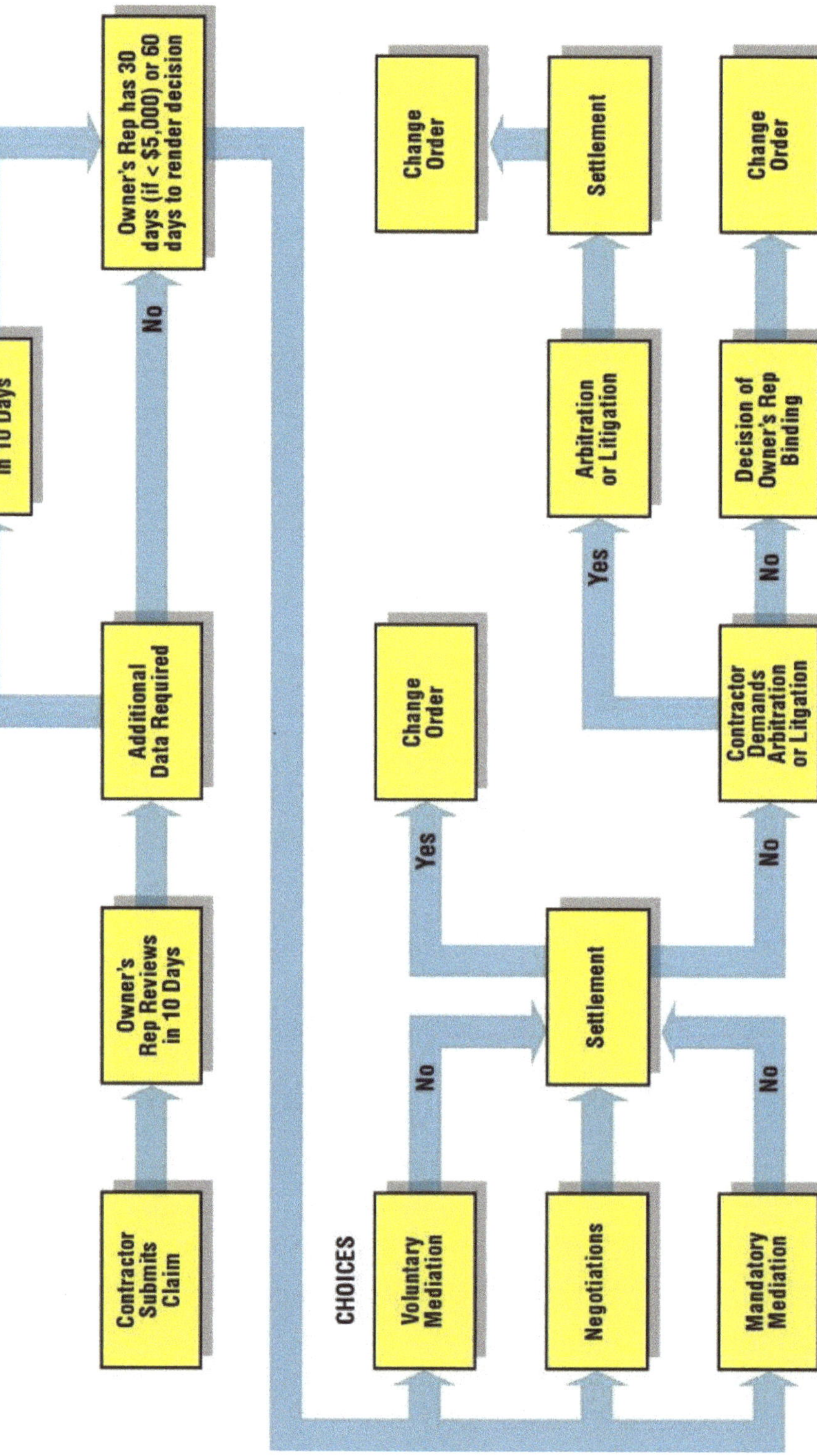

Chapter 11

Cost Control for the Project Manager

Chapter 11

Cost Control for the Project Manager

Good cost control is the hallmark of successful project managers. How do you do it? If it were easy, everyone would do it well. Well, it is not easy, and the bigger the project, the harder it is to do well. Controlling costs is basically the same whether the project is awarded on a cost-reimbursable or lump-sum basis. The difference is whose money is at risk—the owner's or the contractor's.

The first principle of cost is that every project has an ultimate cost. It can be minimized by clever execution or by removing unneeded scope and features, but there is an ultimate cost. The trick is first to estimate that cost early and then manage to it. Figure 11-1 describes the cost influence curve:

Figure 11-1: Opportunity to Influence Project Expenditures in Construction

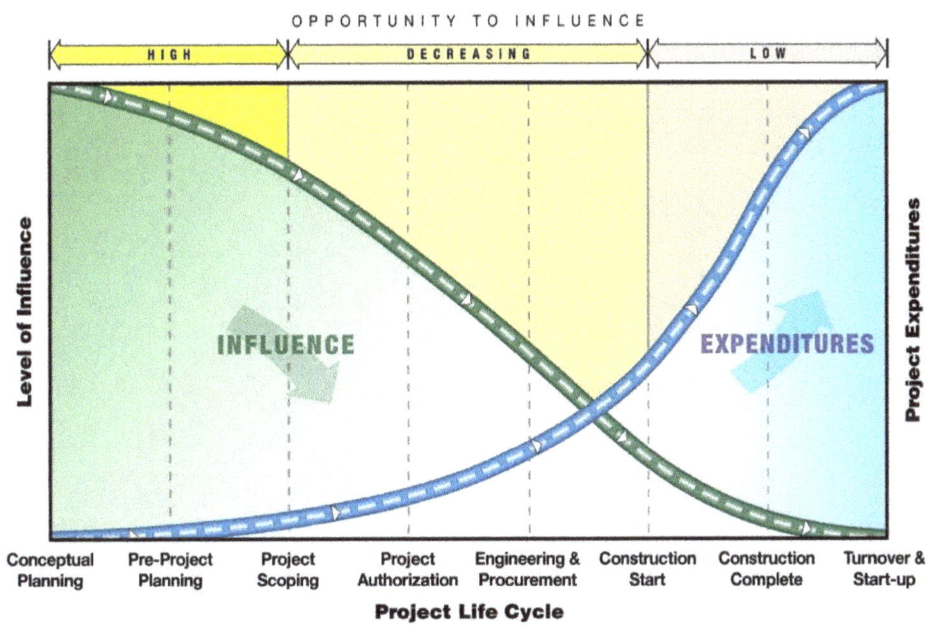

This curve, which has been extensively documented, demonstrates that the cost of a project can most be influenced at the start, during scope definition. As the project proceeds, the ability to influence the final cost decreases so that by the time detailed engineering and procurement are underway, the ability to influence the cost is in rapid decline. By the time engineering is complete and construction is underway, the ability to influence the project cost is very low.

The second principle of cost is that small things can have large effects. It is relatively easy to keep track of the cost of large elements—major equipment, for example. It is quite another to keep track of small things—electrical cable, pipe fittings, or conduit fittings, for example. Designing, procuring, and installing the "little things" (called bulk materials in the process plant industry) can cost up to 60 percent of the total project costs. A 10 percent overrun across the board in bulks could result in up to a six percent increase in overall project costs. Even the practice of overbuying to create an "available surplus" of, say, 10 percent, can increase the overall cost by two to three percent. The little things can really add up.

In this section, the following topics are discussed:

- Definitions
- Estimating
- Execution phase
- Cost overruns
  - Project definition
  - Contracts
  - Owner
  - Schedule
  - Cost of home office and engineering services
  - Procurement and subcontracting
  - Construction
  - Joint venture management
  - Other important points
  - Cumulative impact of change

11.1 Definitions

Cost control is defined below:

- **Deviation:** Any change from the current plan, whether it has cost, schedule, or quality implications. Deviations can be mitigated so that there is no impact on cost, schedule, or quality, or they can become trends or change orders.
- **Trend:** A deviation that has been identified to have cost, schedule, or quality implications. On cost-reimbursable contracts, trends are increases in the cost or extensions to the project schedule and are formally transmitted to the owner. On lump-sum projects, trends that do not become change

orders are additional costs to the contractor and decrease its profit on the project.

- **Scope change:** A deviation that includes a change in project scope from the contractual basis.
- **Change order:** A trend or scope change that has cost, schedule, or quality implications and that changes the contractual basis for the project—extending the schedule, increasing the base cost, or changing the delivered project. Normally both the contractor and the owner formally sign off on change orders.

11.2 Estimating

It is beyond the scope of this book to describe the art of estimating in any detail. Nevertheless, a good estimate (not fat and not un-creditably low) is the key to successful project execution. Knowing how to obtain a good estimate can be the secret to a good project.

It is often said, "There aren't any good estimators anymore. They all retired, quit, or got promoted." That cannot be true. Projects today are more complex, have more participants and, in some cases, are larger. Good estimates keep being produced. Companies have developed cost databases that can provide historical cost information from which to complete the estimate. The trick is knowing what is in the database and how to apply it. Here, a good estimator is worth his or her weight in gold, but a good project manager working closely with estimating can accomplish the same result.

The project manager must deeply involve him or herself in the estimating process. That does not mean doing the estimate (except perhaps for a small project). It means being involved in every decision, checking the scope that is being estimated, and testing the estimate using rules of thumb.

Above all else, the project manager must be the ultimate manager of the scope of an estimate. Deciding and documenting what is to be included in the estimate can save a lot of heartache and rework later. In creating the estimate scope, a lot of the uncertainties in the project will surface and can, hopefully, be resolved (or at least an allowance or contingency can be added). It is easy to let the estimators work up the estimate with the engineers, procurement people, and construction staff, but by not involving him or herself in the day-to-day decision making, the project manager risks receiving an estimate that does not really represent the project or (often) is high and no one knows why.

Once the estimate has been prepared and has undergone the normal checking, the project manager (and often his management) needs to review the estimate thoroughly. The first level of review should be with the engineers and others who created the estimate input. Check that the correct scope was used, that no additional contingency was added during this phase, and that every good cost-saving idea has been developed and included. This is also a good time to check the man-hour estimates of engineering and other home office personnel. Some of the best reviews require each engineering and home office discipline to justify its man-hour estimates through discussing the scope it will perform and comparing the project to historical norms for similar projects. These reviews are typically group sessions with all disciplines present to look for overlaps or gaps in scope among discipline groups or differences in approach. Spending half a day to several days on this detailed review is normally time well spent. By the end of the review, the project manager should have a firm grasp of the details of the estimate basis.

11.2.1 Estimate Types

In the process industry, there are several different types of cost estimates. In many reimbursable projects, three estimates are made during the project. Near the beginning of the project, estimators, working with the process engineers, make an estimate based on the capacity of the process units and historical costs for similar units, adjusting the capacity by an appropriate factoring technique. This is referred to as a capacity factored estimate and is often described as having ± 25 percent accuracy.

As greater design work is completed and pricing for major equipment is available, a materials and equipment factored estimate is prepared. This estimate, based on the major equipment and materials prices, uses historical relationships between the costs of major equipment and materials to arrive at the total project price and often has ± 10-15 percent accuracy.

Finally, there is the definitive estimate, also sometimes called a detailed estimate. This estimate is based on full details of the project—actual pricing of equipment and materials, detailed takeoffs of all materials, complete estimates of engineering and home office services, and detailed construction and staffing plans. Definitive estimates often are considered to have ± five percent accuracy.

11.3 Execution Phase

During the execution of the project, the project manager should take responsibility for various elements of cost control. With the assistance of the project cost controls team, the project manager should:

- Develop an overall project control plan that gives advance warning of undesirable trends, deviations, slippages, and other project problems to facilitate timely corrective action to minimize the impact on project cost, schedule, and quality.

- Design the measurement and control tools including the automation plan to be used on the project.
- Obtain owner approval, when required, for the Cost Control plan and tools to be used.
- Review and approve the project cost code of accounts and work breakdown structure.
- Ensure that exception reporting is implemented to keep project management team members informed of deviations from the project execution plan.
- Ensure that cost trending and change order procedures are implemented early in the Execution Phase and that timely reports are available to involved project participants.
- Lead the preparation of and approve original project budgets and subsequent revisions.
- Ensure that cash flow forecasts are made and monitored.
- Lead the risk analysis to set project contingency.
- Control the use of project contingencies.
- Initiate and approve financial reports to the owner and contractor management (normally monthly).

11.4 Cost Overruns

Nearly all cost overruns originate from one of the following causes, which are each described below:

- Poor project definition
- Poorly drafted contract
- Actions of the owner
- Unrealistic schedule
- High cost of home office services
- Overruns in procurement and subcontracting
- Construction productivity
- Joint venture management
- Joint venture agreement
- Other important points

11.4.1 Project Definition

No one normally starts a project with the notion that the project is poorly defined. Poor project definition is often discovered later when comparing the "as-built" scope to the "as-estimated" scope. The project manager must conduct a thorough review of the project scope at the start and agree on it in detail with the client. If the agreed scope is not what was estimated or priced in the contract, a change order must be prepared at the very start. It is amazing how often this step is overlooked in the excitement of project kickoff and how often, later, there is no agreement between the owner and the contractor about what the original scope of the contract was.

Another source of cost overrun in project definition is not having the basic process set (for a process plant). Many owners ask the contractor to provide "one each—process 'X' plant." The process plant can be supplied in a variety of ways, and generally the owner will have a lot of input. The contractor who commits to providing "one each—process 'X' plant" is at the owner's mercy, and a very adversarial relationship is sure to ensue. Take the time, through Front-End Loading, to define the plant thoroughly.

Another source of poor project definition is unclear, inapplicable, or incomplete project conditions and specifications. Often owner engineering organizations will specify that their (multivolume) engineering specifications should be used for a project when they may not apply directly or may need modification. Later, the owner engineering organization may insist that changes to the specifications to make them applicable are at the contractor's expense or that the contractor must conform, even though they do not apply exactly.

The project manager must also watch out for subtle changes in conditions—geographic, climatic, seismic, or soil conditions. These may not have been sufficiently defined at the start of the project or may be discovered to be different from those specified in the contract. Likewise, administrative requirements of the owner or a governmental agency may be quite different from what was anticipated at project start. Local labor rules, requirements for local content, and local vendors supplying materials may cause project costs to change. In particular, import regulations and taxes are often a source of cost "surprises" to the project and should be reviewed thoroughly at the start of the project and monitored constantly throughout the project for unanticipated differences.

11.4.2 Contracts

Contracts are normally negotiated at the start of a project, often under severe time pressure. Not only can every eventuality not be anticipated at the start of the project, but contract drafting defects can creep in. Contract clauses can be vague, ambiguous, or one-sided. Conditions for contract completion can be incomplete or unclear, particularly in the areas of transfer of care, custody, and control and startup and performance tests. Warranties, how they are enforced, the dates and conditions for expiration of warranties, and the contractor's responsibilities for warranty "make good" are often neglected. Final contract closeout, including obtaining completion certificates, collections of reserves and retainage, conditions to be met to receive final payments, and release of bank bonds, need to be drafted into the contract and monitored as to applicability at the end of the contract.

Agreements with potential subcontractors and other third parties are often not finalized at the effective date of the contract, creating potential for extra cost. The project manager must monitor subcontracting to be sure that subcontract Agreements have terms and conditions consistent with the prime contract and that there are no overlaps or gaps in the prime contract versus subcontracts.

11.4.3 Owner

Actions the owner takes can also cause cost overruns. Inexperienced owner project management team members (inexperienced in projects in general or with the specific project type) often cause increased costs. Delays in owner decisions or late modifications can also increase costs through reduced productivity and necessary rework. If the owner's preferred vendor and subcontractor list is too limited, or if unsuitable suppliers are included, poor-quality materials or higher costs for the materials may be incurred.

A project several years ago required that the owner have 15 calendar days to approve certain drawings. After that, if no comments had been received, the contract permitted the contractor to treat the drawings as owner approved. In actuality, the owner *did* submit comments on the drawings, but on the average, they took 25 days. In the meantime, the engineers and designers had gone on with their work and subsequently had to go back and incorporate the owner's comments (to keep peace with the owner although the contractor was not contractually obligated to do so). In the end, the owner's tardy comments cost at least a month of overall schedule slippage and many millions of dollars in engineering and construction rework. Could the project manager have avoided this problem? Probably—by taking a stand on comments and the time the owner had to make them.

11.4.4 Schedule

The importance of a 90-day schedule and a detailed execution schedule created at the very start of a project cannot be overemphasized. Costs can be increased either through unrealistic schedules being created or imposed (by management or by the owner) or through not meeting the agreed schedule. Areas of poor schedule performance that can contribute to increased costs include:

- Late mobilization of personnel and resources both in the engineering office and the field;

- Decisions not being made at the time specified by the schedule;
- Performing activities out of sequence (*e.g.*, mobilization of subcontractors too early);
- Late revision of documents considered finished;
- Delays in releasing purchase orders to vendors;
- Vendor drawings and data being issued late to the design team;
- Delays in issuing documents to the vendors and to the job site;
- Late deliveries of materials to the job site, in particular, bulk materials; and
- Cascading effects of delays of one discipline on another on the job site.

11.4.5 Cost of Home Office and Engineering Services

Home office and engineering services costs can be overrun by a variety of causes, such as:

- Internal productivity problems caused by:
    - Performing work out of sequence, resulting in increased costs from affected disciplines or rework;
    - Uncontrolled interfaces among disciplines; and
    - Impact of changes by one discipline on the work of another;
- Late modifications and repetition of studies;
- Excessive management reports and documents;
- Excessive costs for computers and information technology; and
- Underestimating the man-hours required for job site support.

11.4.6 Procurement and Subcontracting

Even the best procurement and subcontracting plans can run into cost difficulties. The market conditions for purchasing and subcontracting can be different from those anticipated during estimating, with inflation or lack of qualified and available suppliers causing increased costs. The workload of selected vendors can cause increasing prices and delay in delivery. If the purchase order documents are too voluminous or poorly structured or the terms and conditions or specifications too strict or not complete enough (*e.g.*, no agreed conditions for variations), the vendor can suffer increased costs that it will try to pass along. Spare parts not identified at the time of purchase will not normally be competitively priced after the purchase order has been placed. If any purchasing or subcontracting constraints are imposed by project financing and are not properly accounted for in the estimate, costs can increase.

Finally, a favorite subject in managing costs: bulk materials. Bulk materials are generally purchased by providing bidders with takeoffs of the expected quantities and types. If the quantities used for selecting the vendor do not represent final requirements, even the most rigorous purchasing strategy may result in selection of the wrong (not lowest cost) vendor. If the types of materials given to the bidders do not include all requirements, the supplier may provide the additional types of materials at a noncompetitive price.

11.4.7 Construction

Increased costs can be incurred during construction from several causes. The impacts of these increased costs can be passed along to the owner, the general contractor, or others depending on the contractual relationships. The project manager should be aware of the impacts and take the steps necessary to mitigate them.

Construction costs can increase because of poorly qualified or experienced construction supervision, slow manpower mobilization, and delays in initial site work, which can impact succeeding construction activities such as underground piping, slabs, foundations, concrete structures, and structural steel. Delays in delivery of drawings and materials can cause lower productivity, schedule delays, and rework. Poor shop inspection of fabricated equipment and materials can cause the field to have to rework the items or return them to the fabricator. Late modifications of construction documents or late recognition of pre-commissioning and commissioning requirements can cause rework and increase costs. Failure of subcontractors to perform as planned or the financial failure of a subcontractor can lead to increased costs and potentially large delays in construction performance.

11.4.8 Joint Venture Management

Joint ventures, either of owners or contractors, can lead to cost overruns from a variety of causes. Managing the interfaces in a joint venture can become a full-time job for the project manager, who may need to add staff to monitor and manage them. Having a well-qualified project management team to assure technical coordination (choice of and compliance with rules, standards, and specifications), coordination of technical interfaces, and control of all joint venture expenditures, *at time of their commitment*, can make the joint venture a success.

Common sources of overruns in joint ventures are:

- Enforcing common methods in the joint venture rather than using each partner's normal work processes and tools;

- Coding systems for common materials and documents implemented late or poorly;
- Split of responsibilities or services that are vague and/or unsuitable;
- Lack of motivating provisions for each company to keep their own costs under control and to keep from generating any increased costs for their partners; and
- Invoicing procedures to the joint venture being too complex and costly.

11.4.9 Other Important Points

Other sources of cost overruns in executing a project include:

- Premature withdrawal of a key person from the project management team with the resultant loss of knowledge and inertia; and
- Exchange rate risks.

11.4.10 Cumulative Impact of Change

Change usually increases project cost. But if project change becomes excessive, the "cumulative effect" of change can really add to the cost of the project—much more than the sum of the individual changes. Many studies show that the threshold for significant cumulative effect is about 10 percent of the contract price. When change exceeds 10 percent, in the aggregate, the ability of the project to evaluate and estimate the actual cost of each change diminishes substantially. Change is overlaid with more change, and costs tend to spiral out of control.

The Construction Industry Institute (CII) has done a lot of research on the cumulative impact of change. The following thoughts are from CII Research Publication 43-2, "Quantitative Effects of Project Change:"

- "Projects with a high percentage of change experience lower craft productivity than projects with little or no change."
- "Projects with less than six percent change, on average, experienced better productivity than planned."
- "Productivity rates (work-hours per unit of measure) generally used in the construction industry presume that an average amount of change (six percent change orders) will be experienced."

- "Most estimators take into account an expected inefficiency associated with implementing a change because the work may be out of sequence. However, few consider the effect of implementing multiple changes or large changes on the overall productivity of the project. This research indicates that this impact is significant and should be recognized."
- "The vast majority of projects that were studied required more hours to complete the original engineering scope (absent changes) than originally budgeted."
- "The research found that even with projects experiencing no change, the statistics indicates an average engineering overrun of seven percent."
- Projects with less than five percent engineering change did not experience construction productivity lower than 1.0.
- At 25 percent construction change (in man-hours), construction productivity is about 0.93 compared to 1.02 when no change occurred.
- At 25 percent engineering change (in man-hours), engineering productivity is about 0.88 compared to 0.93 with no change.
- At 25 percent engineering change, construction productivity is 0.9 compared to 1.02 with no engineering change.
- "Construction productivity declines significantly with increasing engineering change."
- "Projects that start well and that are executed with little engineering change tend to run well in the field."
- "Projects cannot endure numerous changes without suffering a decline in overall project cost performance."
- "Changes that occur late in a project are implemented less efficiently than changes that occur early in the project."
- "A Project Manager who may be experiencing a high degree of engineering rework and Client requests near the end of a project should seriously consider the impact of these changes on the main scope work. [Experience] would indicate that it may be advisable to defer discretionary changes until after the main scope of work is complete, if possible, or to assign a different crew to the change work. Even if the change is non-discretionary, it may be prudent to schedule implementation during periods when little other change work is anticipated."
- "…construction projects are unique, one-time events. …Yet Project Managers often discount the early trends of a project, believing that time remains to recover schedule losses or negative trends."

- "…projects that are trending late tend to finish late, and the ability to recover schedule declines with time, frequently due to an inability to resist ongoing changes."
- "Very rarely do projects with early forecasts over budget actually come in under budget."[33]

[33] The Construction Industry Institute, Publication 42-3, "Quantitative Effects of Project Change," Austin: Construction Industry Institute, 1995.

Chapter 12

Effective Change Order Management

Chapter 12

Effective Change Order Management[34]

The preparation, estimation, and resolution of change orders can present challenges to any engineering and construction project. The all-too-common practices of ignoring, delaying, or taking a prejudicial approach to the processing and resolution of change orders can potentially lead to distrust between parties while the issues remain unresolved. These inappropriate practices often result in interruptions in the flow of work, increases in project costs, contract milestone delays, and possibly an extended date for project completion.

Change orders can also become unmanageable if the process is not sufficiently defined, or if the individuals involved in the preparation and administration of change orders are inexperienced. As is often the case, claims and disputes occur when change orders are not properly addressed. Therefore, the timely resolution of change order issues is key to a project's success.

The following topics are discussed:

- Change order definition and other change order-related considerations;
- The reasons that a potential change order may be introduced on a project;
- Factors resulting from change orders that influence a project's performance and costs;
- The various methods of calculation that may be used to estimate the cost of a potential change order;
- Procedures for initiating a change order;
- The development of an independent estimate for the owner's[35] use in comparing and validating the contractor's change order cost proposal;

[34] References for Chapter 12: Barrie, Donald S. and Paulson Jr., Boyd C., *Professional Construction Management*, McGraw Hill, Inc., 2nd ed., New York, New York, 1978, 1984; Fisk, Edward, *Construction Project Administration*, John Wiley & Sons, Inc., New York, New York, 1978; The Construction Industry Institute, Project Change Management, Special Publication 43-1, University of Texas, Austin, Texas, November 1994; James M. Montgomery Consulting Engineers, Inc., *JMM Construction Management Manual*, CM 1102 Initiator Change Order Request, CM 1103 Change Order, CM 1106 Deviation Request, CM 1107 Technical Justification, 1987.

[35] The term "owner" herein implies reference to the owner or the owner's representative.

- Review of the contractor's request for an extension to the time for project completion as a result of a change order;
- Successful negotiation and finalization of a change order; and
- Change order recommended practices and procedures.

12.1 Change Order Definition and Other Change Order-Related Considerations

A change order is a document that contractually alters an original agreement between the owner and contractor. Relative to construction, a change order is the result of an owner-approved revision to terms and conditions (*e.g.*, scope of work, performance time frame, price, etc.) that are defined in a project contract.

Comparatively, field work orders (FWO) are directives from an engineer or construction manager that either clarify or request minor changes to the contract documents. FWOs are typically minor revisions that usually do not significantly impact the contract budget or schedule.

Change orders can be either directed by the owner or requested by the contractor. Depending upon its specific requirements, a change may or may not impact a project's contractual cost and schedule. If the cost is affected, then a change may be represented as either an increase or a credit. Similarly, the durations of activities in a project schedule that are identified with a change may be extended or reduced. If these activities are on the schedule's critical path, the project completion date may be impacted.

It is important that the contract completely defines the project scope of work to minimize disagreements over scope changes. However, parties commonly disagree over what scope or factors may comprise a change order and/or its impact. This disagreement is not unusual because the nature of each change will predictably affect each party in different ways.

The resolution of change orders can also become the source of disputed issues between the owner and the contractor. Inadequate management of the change order process can further intensify the disputed issues. The owner's and the contractor's diverging and competing interests relative to the changes may also become a factor in judiciously resolving differences.

The owner's priorities are focused on predicting and controlling costs so that the funding budget for a project is not jeopardized. Also, the owner usually places a high priority on achieving a certain project completion date, and often will include contractual liquidated damages to be paid by the contractor if the contractual completion date is not achieved.

This contractual completion date will need to be adjusted for any delays for which the contractor may be entitled to a time extension.

Change orders often entail several complicating factors. First, a contractor may view change orders as increasing project scope and offering a source of additional revenue. However, this supplemental work can also present the contractor with the problem of having to proceed with the completion of change-related scope of work prior to negotiating and finalizing the entitlement and payment for a change. The contract may require the contractor to proceed with the extra work pending the resolution of the valuation of a change order, which may also require the contractor to finance the cost of the alleged additional work until resolution occurs. The contractor will normally proceed with scope changes if it is confident that it will be fairly compensated for the alleged extra work at a later date. The contractor also has an incentive to complete the contract work in a timely manner because "time equals money," *i.e.*, the cost of the contractor's field staff and other time-related costs continue to rise if the contract work is delayed beyond its planned completion date.

Construction contracts often include conflicting clauses regarding changed work. On one hand, a contract may state that the contractor cannot proceed with changed work without a written and approved change order. On the other hand, contract language may also state that the owner has the right to order the contractor to perform extra work without an agreement for compensation. These conflicting clauses force the contractor to choose between 1) performing the extra work at the risk of possibly later having to accept a reduced payment or no payment, or 2) refusing to proceed with the work. The latter choice may place the contractor in a position of default, which may result in the owner terminating the contractor.

12.2 Factors that Establish a Change Order

A number of different factors can influence the development of change orders. These elements can result from either foreseen or unforeseen conditions and include, but are not limited to, the following:

- **Design modifications** involve revisions to contract drawings and specifications. The owner, owner's representative, or owner's engineer usually initiates these changes. Design modifications are generally the primary cause of change orders on projects.
- **Errors and omissions** address items that are determined to be defective or missing in the contract set of drawings and specifications.
- **Changed conditions** reflect the actual, physical conditions that are encountered but unforeseen and materially differ from the conditions that

were identified in the contract documents, or conditions not normally found in the area of the project site.

- **Additional/reduced work scope** occurs at the owner's discretion when the contract scope of work expands as a result of design enhancements or is reduced because of budget considerations and value engineering.
- **Owner-directed schedule acceleration or slowdown:** is based on the owner's directive to modify the contractor's planned performance on a project, the contractor incurs additional costs and expenses when the scheduled project performance is accelerated or delayed. In addition, if the contractor does not receive justifiable time extensions for excusable delays, it may require that the contractor accelerate the work to mitigate the delays and meet the contractual milestone dates.
- **Work sequencing** results from an owner's requirement to revise the contractor's planned sequence for completing work tasks. The owner's directive may impact the contractor's scheduled and available labor, material, and equipment, and the contract time for completing a project.
- **Adjustments to unit-pricing** are based on cost overruns resulting from revisions to contract unit-prices. Increases in a contract item budgeted quantity or a material item budgeted quantity are the most common reasons for an adjustment to a unit-price.
- ***Force majeure* delays** generally include schedule extensions based on events for which the owner and contractor are neither accountable nor able to control, such as: 1) acts of God, 2) unforeseen labor shortages, 3) strikes and disputes, and 4) unusual weather-related delays.
- **Added cost incentives** are implemented by the owner to encourage the contractor's participation in value engineering. An owner may also make a cost incentive to entice the contractor to accelerate project completion.
- **Delayed, denied, or restricted access to a project site** occurs when a contractor cannot fully access a project site because the owner cannot grant access or obtain permits.
- **Inappropriate rejection of the contractor's work** is based on the owner's incorrect rejection of a contractor's completed work elements. A change order may be appropriate if subsequent investigation indicates that the work completed was acceptable and in accordance with the contract specifications and terms and conditions.
- **Delays to owner-supplied services and materials** may result in impacts to the project schedule that result from an owner's inability to provide services, engineering deliverables, permanent materials, and/or equipment within a timeframe consistent with contract requirements.

12.3 Change Order Backlog

Both the owner and the contractor benefit if potential change orders are processed in a fair, equitable, and timely manner. The failure to do so most often increases the probability of extended disputes and claims between the owner and the contractor.

A significant amount of change order backlog, in terms of quantity and value, indicates that a project is in trouble. Backlogs are typically attributed to a contractor's above-average submission of change order requests and/or the owner's inability or reluctance to promptly process the potential change orders.

A large backlog of unresolved potential change orders generally falls into one of the following categories:

- **Undisputed:** Includes potential change orders that are not in contention. Their processing is delayed because of inadequate administrative support, procedures, or funding.
- **Disputed:** There are two reasons for disputing a potential change order. In the first case, the owner and the contractor cannot agree that the scope of work identified in the potential change order represents a change to the contract scope. In the second case, the owner and the contractor agree that the work scope is a change to the contract scope of work but cannot agree on the value or cost of the potential change order and/or its time impact. If potential change orders are disputed, approval and compensation may be delayed. If ignored or unresolved, the issues can later become claims.

12.4 Factors Resulting from Change Orders that Influence a Project's Performance and Costs

There are numerous elements that can influence the cost of a change order. The work for change orders can necessitate that the contractor redirect the labor, material, and equipment resources that were originally budgeted, scheduled, and designated to be used to support the completion of the tasks identified for the contract scope of work. As a result, the contractor's plan for completing the project may be impacted. The factors from change orders that may affect the planned performance of a project include the following:

- **Original project scope:** Change orders can have a combination of direct, indirect, and cumulative impacts on the cost and schedule to perform the original scope of work. During the later phases of a project, the impacts that result from a change can become even more significant and complex. The reason for this is because during the later stages of a project, construction is in full swing, and there are more parties and elements engaged. It also becomes more difficult and less efficient to implement a change as time progresses on a project.

- **Loss in productivity:** The contractor evaluates how the work for the change will be performed and what effect this added work will have on the original contract scope of work. If the work related to a change places the contractor's labor and equipment resources in competition with the original project scope of work, the performance of project work may be adversely affected. CII notes that "on projects with less than 6 percent of hours spent on changes, productivity was better than planned."[36] Conversely, when more than six percent of planned hours were spent on changes, productivity was worse than planned.
- **Schedule interferences:** The contractor should consider whether the work defined for a change will affect the sequence and duration of the activities in the contract schedule. If scope changes impact the activities on the schedule's critical or near-critical paths, the contract completion date of a project may be extended unless the work is accelerated. Additional work scope may also affect contractual milestone dates. Disruptions to the schedule from a change order may result in additional costs, irrespective of any change to the completion date.
- **Subcontract impacts:** A contractor notifies and shares with its subcontractors the documentation for each potential change order affecting a given subcontractor's work. Based on this information, the contractor instructs its subcontractors to fully evaluate the cost and the schedule impact that a potential change order may have on its contract work scope. This information is then incorporated into the change order documentation that the contractor submits to the owner.
- **Cumulative impact of multiple changes:** The contractor must evaluate the collective impact that change orders have on its resources, including administration, field supervision, labor, equipment, schedule, site congestion, etc. As the number of change orders grows and the performance of work related to change orders becomes more compressed, the contractor's planned productivity for performing the original contract work scope can diminish. Therefore, it is important that the contractor assesses the collective impact that numerous change orders may have as the project progresses. This includes identifying and appraising anticipated cumulative demands that the extra work places on the planned work effort and resources. The failure to do so may require a contractor to file a claim at the end of a project to collect for the cumulative impact from all

[36] Construction Industry Institute, "Project Change Management," Special Publication 43-1, November 1994, p. 7.

approved and disputed change orders. It is also important to note that the impact and effect of the change orders may be more important than the number of changes.

- **Exposed risk:** In assessing a potential change order, contractors should consider any uncertainties that may be associated with the revised work scope. The contractor should add additional cost and time to its cost proposal submittal for a potential change order as a contingency to compensate for the possible impact from these undefinable factors.

12.5 Methods for Change Order Basis of Calculation

The pricing for potential change orders can be developed using one of several methods. The choice of method can be established in a contract, in a post-contract agreement between the owner and the contractor, or at the contractor's discretion. Each method has distinct advantages and disadvantages that depend on the nature of a project and the requirements specific to a given change. Pricing is typically based on one of the following practices:

- **Unit-price:** Based on the unit-price items that are approved as part of a project's construction contract between the owner and the contractor.
- **Lump-sum:** The owner and the contractor agree that the added work scope, which the change defines, will be completed on a fixed-price basis, including the costs of the contractor's overhead and profit. The contractor accepts the risks and liabilities associated with completing the extra work for an agreed-to price when the extra work is completed using this method. To ensure adequate compensation in the contractor's pricing for the proposed change order, the contractor should include additional costs for contingency to offset the risks and unknowns that it perceives to be associated with the added work scope. The estimated costs are then evaluated and negotiated between the owner and the contractor to ensure that the finalized cost for completing the extra work is fair and reasonable.
- **Time and materials (T&M):** In this method, the value of the scope of the extra work that is identified for a potential change order is based on the actual direct expenses for supervision, labor, materials, equipment, etc., that the contractor incurs. Factors are also included that address allowances for the contractor's overhead and profit requirements. In many projects, provisions for the contractor's overhead and profit factors are addressed in the general conditions section of a project's specifications or in the changes section of the contract. When a contractor completes extra work for a potential change order on a T&M basis, the associated risks and liabilities are the owner's responsibility unless the owner can demonstrate that the contractor's costs were unreasonable. The T&M method is often used if

the owner and the contractor cannot agree on the value of the extra work for a potential change order or if the full scope of the change cannot be clearly defined at the time that the change order is identified, and the owner directs the contractor to proceed with completing the work. Also, the contractor may opt to use the T&M method if it prefers not to assume the risks that may be associated with a potential change order.

12.6 Procedures for Initiating a Change Order

A contract typically provides that either the owner or the contractor may introduce a change order. The procedures for initiating a change order tend to differ among projects, as owners and construction managers each have their own established requirements and preferences. Although there may be some variances, the following items address the practices normally expected on a project.

- **Deviation request** (Figure 12-1): The contractor completes a deviation request form and submits it to the owner or its representative. The purpose of this form is to document the deviations that have been determined from the original contract documents. This form serves as the contractor's notification of any changed conditions, omissions, or discrepancies that may result in extra work. The owner can either approve or disapprove the request, depending on the validity or justification of the change.

- **Initiator change order request** (Figure 12-2): The owner, the owner's representative, a project engineer, or the contractor can initiate this form. The preparation and submittal of this form are intended to define the scope of work involved in completing a potential change and to demonstrate the justification for the change. The owner's signed endorsement indicates its authorization to proceed with the issuance of a change order. The change order request offers notification that the possibility exists for adjustments to the contract price and schedule because of a potential change order.

- **Change order technical justification** (Figure 12-3): The owner's representative or a project engineer usually prepares this form, which is intended to provide the owner with documented credibility and justification relative to completing the work scope for a potential change order.

Figure 12-1: Deviation Request

LONG INTERNATIONAL

DEVIATION REQUEST

No. ______

Owner ______

Contractor ______

Project ______

Project No. ______

Date ______

Subject ______

Drawing Nos. ______

Specification Nos. ______

Original Contract Requirements ______

Reason for Deviation Request ______

Proposed Deviation ______

Any Change to Contract Cost? ☐ Yes ☐ No If Yes, Describe ______

Any Change to Contract Time? ☐ Yes ☐ No If Yes, Describe ______

☐ APPROVED ☐ DISAPPROVED

CONTRACTOR	DATE	OWNER/OWNER'S REPRESENTATIVE	DATE

CM 2020 FORM 01 a.1 13FEB2021

Figure 12-2: Initiator Change Order Request

LONG INTERNATIONAL

INITIATOR CHANGE ORDER REQUEST

No. ______

Project ______________________________

Project No. ____________ Contract No. ____________ Contract Date ____________

Contractor ______________________________

Proposed/Submitted By ☐ Owner ☐ Owner's Rep ☐ Engineer ☐ Contractor

Name ______________________________ Date ____________

Change Order Justification ______________________________

Description of Work to Be Performed ______________________________

Contractor is authorized to proceed with this Change ☐ Yes ☐ No On ____________ DATE

Approved ______________________________ OWNER ____________ DATE

CM 2020 FORM 01 p2 13FEB2021

Figure 12-3: Change Order Technical Justification

LONG INTERNATIONAL

CHANGE ORDER TECHNICAL JUSTIFICATION

Potential Change Order No. ______

Contract Change Order No. ______

Contractor ______

Contract No. ______

Prepared / Submitted By ☐ Owner's Representative ☐ Engineer

Name ______ Date ______

Description of Change ______

Reason for Change ______

Alternatives Considered ______

Impact of Non-Incorporation ______

Approved ______ OWNER ______ DATE

CM 2020 FORM 01 p3 13FEB2021

An important step in the initiation process is for the originator to classify the change as required or elective, with the owner confirming and approving this designation. The determination of whether a change is required or elective will influence the justification for the change and how the change will be evaluated. The differences between the two classifications include the following:

- Required changes *must* be implemented and typically include changes that are necessary to meet the following:[37]
  - the basic, defined venture/business objectives;
  - regulatory or legal requirements; and/or
  - defined safety and engineering standards.
- Elective changes are proposed to enhance a project but are not required to meet the original project objectives. Therefore, elective changes may or may not be implemented. This type of change is not mandated.[38]

The owner's authorization of change order initiation communicates to all parties that the justification and documentation required in advance of implementing a change order have been met.

12.7 Evaluation and Verification of Costs

The owner must have confidence that the contractor's assessment of the costs to complete the extra work for a potential change order is fair and reasonable. There are several actions that the owner can take to avoid being overcharged. The owner's approach in reviewing the proposed costs for a potential change order depends on the contractor's method of calculation that the owner agreed to in preparing the estimate. How the costs are reviewed and verified is determined by whether the costs were based on a unit-price, lump-sum, or T&M type of estimate. The requirements for evaluating estimates that are based on one of these three methods are discussed below.

12.7.1 Unit-Price

In many construction bid forms, the owner includes a listing of unit-price items that the contractor completes by filling in the rate for each item as part of its bid proposal submittal. These unit-price rates typically include the total estimate of direct and indirect costs, including the contractor's overhead, profit, contingency, etc. Therefore, the construction contract between the owner and the contractor includes the unit prices that

---

[37] Construction Industry Institute, "Project Change Management," Special Publication 43-1, November 1994, p. 15.

[38] *Id.*

the awarded contractor proposed. To validate the costs of a proposed change order, the owner needs to verify the material quantities that the contractor reports for the extra work and compare the unit prices that the contractor has applied with the unit prices in the contract. The proposed change order unit prices should be the same as the unit prices in the contract.

12.7.2 Lump-Sum

The owner should prepare an independent cost estimate of the scope of work for each potential change order. This cost estimate should be detailed and comprehensive in format because it is used as the basis for comparing and validating the material quantities, labor production rates, and pricing that the contractor proposes in the submittal for the potential change order.

If possible, the independent cost estimate should be completed in advance of receiving the contractor's potential change order pricing so that a review can be completed without delay.

The owner's independent cost estimate should be based on current market rates and costs as well as those rates and costs that may be specifically established in the contract documents. It is important that the component rates and costs be fair, representative of the work, and defensible. The direct and indirect costs should consider the following:

- **Wage rates:** If a project is supported by union craft labor, the current hourly rates are generally available for the location specific to a project in a published format or on the Internet. Often, the contractor will also provide a copy of the craft labor rates used on a project upon request. The rates that are used in developing the independent estimate need to include payroll burdens, fringe benefits, taxes, and insurance. If it is apparent that these rates do not include those provisions, the owner can add a factor of approximately thirty percent. However, the owner must also consider other elements that may be required for a project and may increase the cost of labor, such as work schedule (*i.e.*, overtime work), per diem and transportation allowances, etc.
- **Material costs:** For its estimate, the contractor will generally base the unit costs of permanent materials on actual invoiced costs or quoted costs solicited from material vendors and suppliers. This information, however, is often not made available to the owner. The owner can either directly contact vendors or suppliers and request this information or base unit material costs on current rates published in a cost guide, such as *R.S. Means, Lee Saylor Construction Cost Index, Page, Richardson Process Plant Construction Estimating Standards,* etc. If required, the owner should also consider and

appropriately account for additional allowances for handling, shipping, storage, tariffs, and duties.

- **Equipment costs:** The contractor will commonly use its corporate rates for owned equipment or rental rates on leased equipment as the basis for estimated equipment costs. The owner usually will not have access to the contractor's rates unless the contract includes them. As an alternative, the owner can directly contact equipment rental firms and solicit the rates. Also, estimating publications such as *R.S. Means* address the average monthly, weekly, daily, and hourly rental rates and hourly operating costs for many equipment items.

- **Subcontract costs:** If a subcontractor is to complete the work or part of the work for a proposed change order, the contractor will often attach a copy of the subcontractor's quotation or estimate. The contractor applies an additional percentage to the price to cover its costs for supervision, administration, and profit. If the extra work delays the contract schedule, the contractor may also request additional compensation to cover the extended overhead on a project. The owner's approach to evaluating a subcontractor's costs for labor, materials, equipment, overhead, and profit should follow the same process it uses in verifying the reasonableness of the contractor's costs.

- **Small tools and consumables:** The contractor normally includes an allowance to offset the replacement cost for small tools and for the purchase of construction supplies and consumable items. This allowance is most often calculated as a percentage of the total direct labor cost, or on a cost per man-hour basis.

- **Overhead:** To cover the costs of the contractor's home office and field management and overhead, it will include a cost allowance based on the estimated financial requirements for corporate and field offices. The percentage, as a factor of the total direct cost for the extra work, varies among contractors and projects. Percentages between eight and fifteen are usually considered reasonable.

- **Profit:** The contractor's profit, as a percentage of the sum of the total direct and indirect costs, should be within a reasonable range. Typically, the contractor's profit on change order work is not expected to exceed ten percent and could be lower.

12.7.3 Time and Materials

For extra work completed on a T&M basis, the contractor's unit rates and markups for overhead and profit that are applied to its actual costs for change orders are based on either 1) the rates that were solicited during the bidding phase of a project and included as part of the project's construction contract between the owner and the contractor, or 2) the unit pricing and markups that were negotiated between the owner and the contractor following the award of a construction contract.

At the completion of the work for a change order, the contractor will submit copies of daily work reports for the extra work to the owner as part of the contractor's monthly payment request submittal. These reports document the actual labor, material, and equipment resources that the contractor expended on an hourly or daily basis[39] to complete the extra work.

In addition to daily work reports, the owner can request that the contractor submit, as backup, copies of the invoices from its subcontractors, material vendors, and suppliers that supported the work.

12.8 Evaluation of Time Impact

The work involved in completing a change order may impact or delay the contract schedule. If the extra work is determined to be non-critical because of the absorption of the total float on the affected activities, the completion date of the contract schedule will remain unchanged. However, if the changed work extends or delays the completion of activities that are on the schedule's critical path, the completion date of a project will slip from the planned date. When evaluating the potential time impact of change orders, the following should be considered.

- It is important for the owner to complete an independent analysis of the project contract schedule to identify which activities the proposed change order work affects and to verify whether the scope of work for a potential change order impacts the schedule and the contract completion date.
- Consideration should also be given to the cumulative impact of multiple change orders on the contract schedule, which could affect the labor productivity of the unchanged work and delay its completion.
- The timely identification of potential issues and delays on projects permits the development of recovery schedules that may help mitigate some or all potential delay.
- The owner should consider including a procedure in the contract scheduling specification that defines the procedures relative to the

[39] Note that the contractor's documentation of the quantities and costs for the labor and equipment resources can be based on hours or days and depends upon the actual duration of the work for the change. For example, if the work took less than a day to complete, then the time and costs for the change would be summarized in hours. If the work took a minimum of one day's time to complete, the time and costs would be expected to be summarized in days.

preparation, format, and submittal of the contractor's schedule analysis for potential change orders. The procedure should also address the approval process and the incorporation of the ratified revisions into the contract schedule.

12.9 Change Order Negotiations and Resolution

The method by which the change order was priced usually dictates the level of negotiation that may be required to resolve additional costs and time. If the change order work proceeded on a unit cost basis, the unit rates used were probably established on the project and agreed to prior to the start of the work for the change. Negotiations for this type of change then focus on the quantities that the contractor reported it used and whether the change warranted an extension of time to the project schedule.

Similarly, for change order work to be completed on a time and materials basis, the unit hourly rates for labor and equipment for the project were probably ascertained in advance of the start of work for the change order. Therefore, the owner bases its approval of the costs for the change order on a review of time card records and copies of material invoices presented by the contractor and, potentially, subcontractor invoices. The project schedule is also evaluated to determine whether the change order work may have affected the completion of critical milestone dates or the overall project completion date.

If the proposed change order was priced on a lump-sum basis, the contractor's estimate proposal includes the risk relative to completing the scope of work for the change. The contractor may also consider that the estimate details for the cost of the proposed change order are proprietary. For this change order method of cost calculation, the elements for negotiations and resolution include the following considerations:

- It is helpful if the owner prepares an independent cost estimate and a schedule analysis of the change order work scope. These help to familiarize the owner with the specific work scope requirements and issues relative to a potential change order and provide a basis for comparison from which to better evaluate and negotiate the contractor's change proposal.
- If timing allows, the cost and time impact of a potential change order should be established and settled prior to starting the change order work. Due to the speed at which projects progress and the urgency of completing change order work, evaluating estimated cost and time impacts in advance is not always possible. When a contractor prepares its proposal during or following the completion of change order work, the estimate is generally based on its accounting of *actual* man-hours expended, quantities of materials used, and operating hours for the construction equipment items utilized. Because the estimate relies on the contractor's records, the owner's

validation can be more difficult than completing an evaluation of a proposal that was prepared in advance of proceeding with the work. The reason for this difficulty is that the contractor's records may not be made available to the owner if such costs are comingled with the base scope of work, which is paid for on a lump-sum or fixed-price basis. If the records are made available, they are often not as complete or comprehensive as they should be.

- For negotiations to succeed, it is important that the owner and the contractor be objective in their analysis of the cost and time to complete the change order work scope. Frequently, the process suffers because of differing personalities and heightened emotions. To be productive, the negotiations should remain focused on the factual circumstances related to the change order.
- During negotiations, there are often disagreements which can impede progress in finalizing a change order. It is important to actively resolve any disputes as they develop so the parties can reach an accord in a timely manner. This attention by the parties will help minimize the filing of claims.

12.10 Finalization of the Change Order

Following the negotiation and resolution of the cost and time elements of a potential change order, the scope that is defined by the change order needs to be fully documented and reported. Typically, change order documentation is captured in the following formats.

12.10.1 Change Order and Change Order Log

Figure 12-4 is a change order document and is prepared by the owner in conjunction with the contractor. This document contains the complete and final language that defines the total cost or credit impact, in addition to any schedule revisions that are related to the change in the contract scope of work. The endorsement of the change order contractually binds all involved parties.

The owner often prepares and updates the change order log (Figure 12-5), but the contractor could be responsible for preparing and updating it with the owner's review and approval. The purpose of the log is to provide a current summary of a project's change order activity.

12.10.2 Monthly Project Schedule Updates

The construction contract typically requires the contractor to prepare and update the contract schedule monthly. In addition to updating the status of activities that are in progress or completed during the reporting period, the contractor addresses in the contract schedule the details of the logic revisions and impacts that the owner and contractor agreed to as part of the change order negotiations. It is important that the owner reviews the contractor's monthly schedule update submittal to ensure that the revisions that were agreed to in the change order have been correctly identified and implemented.

12.10.3 Record or As-Built Drawings

As part of its contract obligations, the contractor typically marks up contract drawings with the as-built details for a project. This task also includes the revisions that were addressed and approved in the change order documents. The owner should periodically review the level of detail and progress of the contractor's as-built drawings. Any deficiencies should be immediately addressed.

Figure 12-4: Change Order

LONG INTERNATIONAL

CHANGE ORDER

Page ___ of ___

Project No. ______	Original Contract Amount	$ ______	Cal Days ____
Contractor ______	Previous Approved Changes	$ ______	Cal Days ____
Contract No. ______	Total Amount of this Change	$ ______	Cal Days ____
Contract Change Order No. ______	Revised Contract Amount	$ ______	Cal Days ____

This Change Order covers the changes to the subject contract as described herein. The Contractor shall construct and furnish the labor, materials, equipment that is necessary or required to complete the Change Order Items for a lump sum price agreed between the Contractor and ______________________, otherwise referred to as the Owner.

Item	Description of Changes	Increase in Contract Amount ($)	Decrease in Contract Amount ($)	Contract Time Extension (Cal. Days)
Totals		$	$	
Net Change in Contract Amount (Increase or Decrease)			$	

The amount of the Contract will be increased/decreased by the sum of $__________ and the contract time shall be extended by____ calendar days. The undersigned Contractor approves the foregoing Change Order as to the changes, if any, in the contract price specified for each item, including any and all supervision costs and other miscellaneous costs relating to the change work, as to the extension of time allowed, if any, for completion of the entire work of said Change Order. The Contractor agrees to furnish all labor, materials and equipment and perform all other necessary work, inclusive of that directly or indirectly related to the approved time extension, required to complete the Change Order items. This document will become a supplement of the Contract and all provisions will apply hereto. It is understood that the Change Order shall be effective when approved by the Owner.

Recommended ______________________ ENGINEER ________ DATE

Accepted ______________________ CONTRACTOR ________ DATE

Approved ______________________ OWNER ________ DATE

CM 2020 FORM 01 p.4 10FEB2021

Figure 12-5: Change Order Log

LONG INTERNATIONAL

CHANGE ORDER LOG

Project No. ______ Project ______ Contract No. ______ Contractor ______ Page ___ of ___

C.O. No.	Description	Dates			Cost/(Credit)	Cumulative Amount to Date	Time (Calendar Days)	
		Initiated	Negotiated	Finalized			Extension	Adjusted Total
					$	$		

CM 2020 FORM 01 p.5 15FEB2021

12.11 Change Order Procedures and Recommended Practices

Changes often have a major impact on project success. The procedures that are defined for change management in the general conditions sections of most construction contracts address the steps to take following the initiation of a change order. However, these guidelines do not normally identify the requirements for managing the change and its impacts. Items that should be considered as part of a project's "Change Order Procedure" include the following:

For the design and preconstruction phases of a project:

- Written contracts, agreements, procedures, and roles and responsibilities should be established and accepted during a project's early phases. If these documents are not in place, difficulties and issues may arise that may affect the execution of the work.
- Define and establish a detailed change management process that is part of the design plan and procedures for the project.
- Anticipate that changes will inevitably occur on a project and plan for them accordingly. If the necessary steps for managing changes are not taken, the change order process will not be managed well, and the project will suffer.
- Attentively monitor the trending of the project's design progress. At the completion of the work related to the design, it is important that the owner initiates and communicates a "freeze" of the design. Any revisions to the established scope of work will be measured against this baseline for the design.
- During the design and preconstruction phases of the project, identify areas of uncertainty and areas lacking fully defined scope. Evaluate the risks that may be associated with these areas.
- Conduct constructability reviews and institute a value engineering program to assess the feasibility of the project's design and defined work scope.
- Introduce changes as early as possible during a project to help minimize the impacts of the changes. Comparatively, it is more cost effective to include changes early in a project than to wait until its later stages.
- Fully define the work scope for each change in terms that all parties understand. A clear and concise definition helps to reduce confusion or the misinterpretation of the requirements of a change by the parties.
- Avoid delaying the approval of changes that will ultimately be approved. This tactic results in the changes being pushed to later in the project, when

they become more costly to implement. It is a misconception to believe that the issues related to change orders resolve themselves.

- Authorization for a change should be mandatory before implementation, regardless of the type of agreement or contract. Authorization should be timely and decisive.[40]
- In advance of releasing the project design and bid package documents for tender and construction, ensure that the documents are complete and fully definitive.

For the construction phase of a project:

- Evaluate the contractor's bid proposal to ensure that the documents are complete and that the cost and schedule reasonably represent the scope objectives and timelines required for the project.
- Review the contractor's construction procedures, methods, and resources to confirm the contractor's capabilities and flexibility to handle the types, magnitudes, and quantities of change orders that are expected for the project.
- Establish a comprehensive system for budget and schedule baseline control.
- Process, approve, and execute change orders in a timely manner to avoid disputes and claims that may affect project progress and completion.
- Do not proceed with the work for change orders without authorization; avoid verbal authorization.
- To avoid confusion and misinterpretations of the work scope and requirements that are defined for change orders, work to ensure that the packages for each change are complete. Actively follow up with the parties involved and solicit any questions or requests for clarification.
- Correct conditions that are determined to be general causes or reoccurring factors for change orders.
- Collect and document detailed information from the changes so that the data may be used to provide a historical reference on current and future projects.

---

[40] Construction Industry Institute, "Project Change Management," Special Publication 43-1, November 1994, p. 29.

12.12 The Change Management Process

There is no "one size fits all" process or set of procedures for change orders. Among the many factors that may influence the structure of the change management process for a given project, a few are contract types, management preferences, logistical issues, and the requirements of owners, contractors, and specific projects. Although there are expected differences in the change processes among projects, Figure 12-6 defines the basic elements and process typically expected during the construction phase of the project.

12.13 Summary

Regardless of effort and the best intentions, project planning is an imperfect process. Change orders are inevitable on any construction project. Although the owner can take steps to help mitigate the number of changes, it is impossible to fully eliminate the issues that impact a project's planned cost and schedule performance. There are simply too many unknowns and variables that are outside the control of management on any given project.

Often change orders can become a polarizing factor among parties. The owner's objectives are to minimize the impact of change orders to the greatest extent possible as a measure of control over a project's budget and schedule. Comparatively, the contractor's interests may involve expanding the scope of work beyond that defined by the contract, in addition to boosting its projected revenue. These differing objectives can result in divergent positions and an inability to reach fair and equitable resolutions on changes. If the owner and/or the contractor chooses to ignore or postpone the negotiation and settlement of change orders until the end of a project, the issues that may previously have been solvable have a greater potential of being disputed and becoming claims.

For a project to be successful, it is important for management to commit to processing change orders in an attentive and unbiased manner. It benefits a project to establish a change order process with a defined set of procedures that is best suited to the project requirements, promotes open communications, is objective, and is structured to foster the resolution of change orders in an expeditious manner.

As change orders are negotiated and finalized, the information should be addressed as updates to the project record, including the project budget and schedule. The timely addition of change-related information provides management with a current perspective of the overall condition of a project compared to its plan. This information also helps the parties more accurately forecast the cost and time required to complete project work.

Figure 12-6: Example of a Construction Change Order Procedure

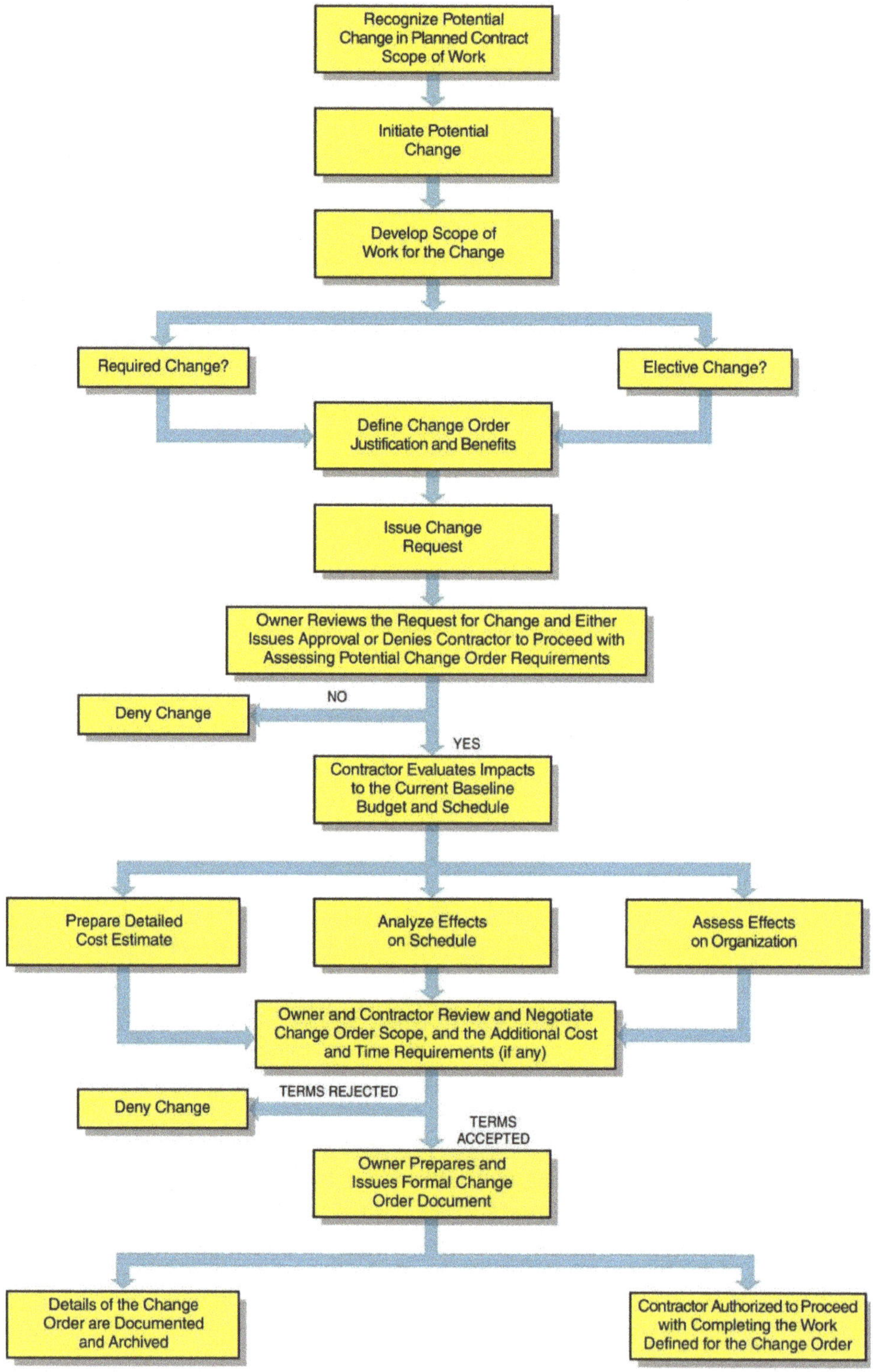

Chapter 13

Keys to Successful Major Project Execution

Chapter 13

Keys to Successful Major Project Execution

For any project to have a successful outcome, there must be consistent application of a suite of effective project management tools. Such application commences from the earliest stage of the project when a project concept and Project Charter are being considered. These tools can be applied to any project, regardless of its size (capital expenditure (CAPEX), or quantum of required human and equipment resources), complexity (process, industrial, commercial, infrastructure), critical success factors, timing (schedule), and governance structure (private sector, government).

In this section, the following topics are discussed:

- What makes a successful project?
- Project initiation and stages
  - Establish a project leader/director
  - Establish the project leadership team
  - Establish the process by which the project life cycle will be managed

13.1 What Makes a Successful Project?

Many organizations define project success as completion of a project within the sanctioned cost and schedule. This is about as rudimentary a definition as possible and is not a good indicator of project success if the CAPEX spend and project schedule have included large contingencies that simply mask underlying project execution inefficiency.

This narrow definition does not encompass Health Safety Security Environment (HSSE) aspects, in particular delivering a *safe* project. Also excluded are the especially important aspects of delivering a project free from engineering (or other) rework (defect elimination). Another missing objective, in the case of a process plant, is for the project to achieve every metric specified in the performance test clauses in the EPC contract, and after successful commissioning, maintain target levels of plant reliability during commercial operation. Also to be considered is whether the project achieved the desired level of transfer of technology, whether the contractor handed over the project documentation in a manner that facilitates easy utilization by operations and, very importantly for a brownfield plant, whether the interface management with operations was handled in an efficient manner and unplanned disruptions to plant operations did not occur. At the conceptual stage of a project, it is important that the project leadership and the owner's management agree on the metrics by which project success will be defined.

In the past years, consensus has emerged that delivery of a successful project requires project management knowledge/experience, skills, tools, and techniques. Project management has always been practiced informally, but in the mid-twentieth century it began to emerge as a distinct profession. The PMI's *A Guide to the Project Management Body of Knowledge (PMBOK® Guide)* identifies its recurring elements, noting that project management processes fall into five groups:

- Initiating
- Planning
- Executing
- Monitoring and controlling
- Closing

Project management knowledge typically focuses on ten broad areas:

1. Integration (management of interface with other stakeholders)
2. Cost
3. Human resources
4. Stakeholder management
5. Scope
6. Quality
7. Schedule
8. Procurement
9. Risk identification (generally HSSE would be included)
10. Communications

This section is based on the author's experiences in the energy sector and examines some guidelines for effective project management control from the owner's perspective. There is one overarching requirement: the entire project execution, from initiation to execution and closeout, must follow a rigid process that ensures the requisite amount of front-end loading (FEL) is applied before, and is in fact a condition for, project sanction. FEL is the process for conceptual development of projects in process industries and may also be referred to as pre-project planning (PPP), front-end engineering design (FEED), feasibility analysis, conceptual planning, programming/ schematic design, or early project planning.

The elemental components of a Stage Gate Process applied to project management are outlined, discussed and, where appropriate, further detailed in this chapter. In a Stage

Gate Process, there are defined activities that must be executed in each stage, and the project is only authorized to progress from one stage to the next with the approval of a "Gatekeeper" who could be an individual or a committee.

13.2 Project Initiation and Stages

Selecting a capable project leader/director, establishing an effective project leadership team, and implementing a Stage Gate Process by which the project life cycle will be managed are key steps to initiating a successful project.

13.2.1 Establish a Project Leader/Director

The project leader/director must have the required level of project management experience for the type of project being contemplated or developed. Familiarity with problems and lessons learned from similar projects are key requirements, enabling thorough examination of the project risk register and the mitigation strategies for identified risks. The project leader/director is the single point accountable (SPA) individual charged with overall project delivery in accordance with the criteria established by the owner and in consonance with a rigorous project initiation, development, execution, and closeout process. The SPA assumes responsibility for the composition of the leadership team chosen to deliver the project.

Some considerations that should influence the selection of a project leader/director include, but are not limited to, the following.

13.2.1.1 Proven Leadership Skills

The SPA ensures that the leadership team functions as a team, effectively leveraging diversity in whatever form such diversity exists, drives alignment of purpose and commitment across the team, and leads by personal example in critical areas such as HSSE drivers. The SPA is responsible for the development of the overall project plan and management of each team member's performance, setting appropriate performance drivers in individual consultations and, thereafter, challenging the team to the highest personal and collective performance standards.

13.2.1.2 Type of Project

Is the project a greenfield or brownfield project? For a brownfield project, effective management of the interface between the executing project team and the operations team is a critical success factor, as very often there are requirements for operating plant shutdowns to facilitate tie-ins. Certainly, no commissioning could succeed without seamless and close coordination of the commissioning with plant operations.

What is the anticipated CAPEX spend? A $2 billion project is likely to be better managed by someone comfortable with such large CAPEX rather than an individual whose largest CAPEX responsibility is $10 million.

13.2.1.3 Project Location

Locations in developing countries, for example, require experience in doing business under conditions that can often be quite different from those in developed countries. A simple example: understanding the importance of the Foreign Corrupt Practice Act (FCPA) and its consideration in setting the Project Charter is vital for a successful project in a developing country.

Another important consideration in executing projects in countries with emerging economies, such as those in Africa, which have recorded significant discoveries of hydrocarbon resources, is familiarity with the host country's policies such as those regarding procurement and local content.

The project leader/director must appreciate the importance of understanding the host government's approval process, in terms of the various stages and the overall timeline for approval, since these could have a major impact on the project schedule, and failure to understand these elements inevitably leads to delays, disruptions, and disputes.

It is also important to have sound knowledge of the availability of capable local human resources, equipment, workshops, etc. to support project execution.

13.2.1.4 In-House vs. Outsourced Project Management

Is the owner outsourcing the overall project management to the EPC contractor on a lump-sum turn-key (LSTK) basis? Will there be an independent project management consultant (PMC) performing all project management functions? In each of the foregoing contractual structures, the experience of the owner's project manager needs to be relevant.

13.2.1.5 Experience in Project Controls

The owner's team must have experience in project controls to ensure that the schedule, change management, earned value measurement, and cost elements of the project are being accurately measured, statused, reported, and controlled. Depending on the type of contract, *i.e.*, lump-sum turn-key, fixed-price, unit-price, or cost-reimbursable, key provisions of construction contracts should be utilized to ensure that identified project risk is properly allocated to the party best able to control that risk. For example, the owner may have contract clauses that reflect no damages for typical delays, such as approval process delays, normal inclement weather, etc. The types of man-hour and cost reports that the contractor would be required to produce to the owner on a periodic basis (usually monthly) are different for the various types of contracts. Labor productivity reporting also is a function of the type of contract being utilized. The owner also needs personnel familiar with the type of scheduling software that the contractor will utilize for critical path method (CPM) schedule development, updating, and control.

13.2.1.6 Experience in Contract Dispute Avoidance and Contract Dispute Resolution

Delays, disruptions, and changes (*e.g.*, engineering changes because of poor pre-FEED and FEED work, hazard and operability reviews, peer reviews, and construction, including unknowns surfacing during site construction) are difficult to prevent. However, an experienced project leader/director and a solid project leadership team can anticipate many delays. To the extent that they identify delays upfront, they can address them with mitigation strategies in the contract documents and an appropriate degree of advanced planning.

13.2.2 Establish the Project Leadership Team

Establishment of the project leadership team is the responsibility of the project leader/director under the appropriate authority vested in him or her by the owner's management. The team members must be chosen based on the design of the project leadership structure.

In a project with no PMC, the project leader/director and the project leadership team assume full responsibility for the traditional PMC role usually adopted on most projects. The project leader/director typically reports to an owner oversight committee or project sponsor. This team undertakes all activities, including direct management of the pre-FEED, FEED, EPC execution phase, project handover, and project closeout. A typical organization structure is shown in Figure 13-1 below.

Figure 13-1: Project Organization Leadership Structure: No PMC—Owner Performs Project Management

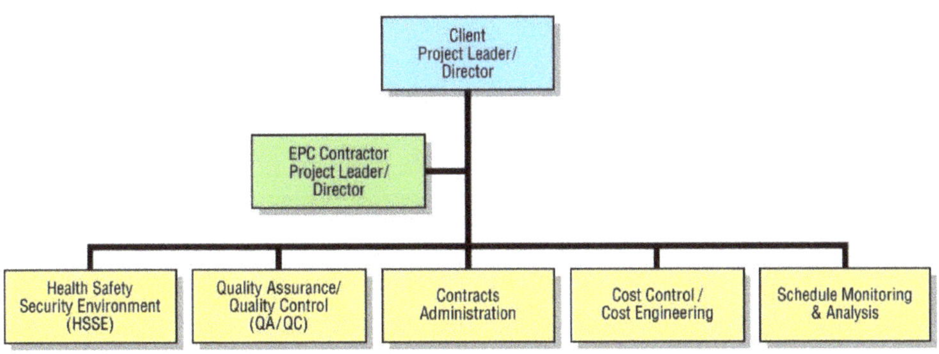

One variation on the above, and other structures that follow, is the insertion of a project controls manager to take direct charge of the activities of the various department leads (HSSE, quality assurance/quality control [QA/QC], etc.) and to collate their activities to produce a monthly progress report.

In the case of the owner outsourcing the project management function to a PMC or to the EPC contractor directly, the project leader/director may not have a leadership team and would be required to act as the owner's representative in addition to assuming responsibility for the overall project delivery. A typical structure for this model is shown in Figure 13-2 below.

Figure 13-2: Project Organization Leadership Structure: PMC Performs Project Management

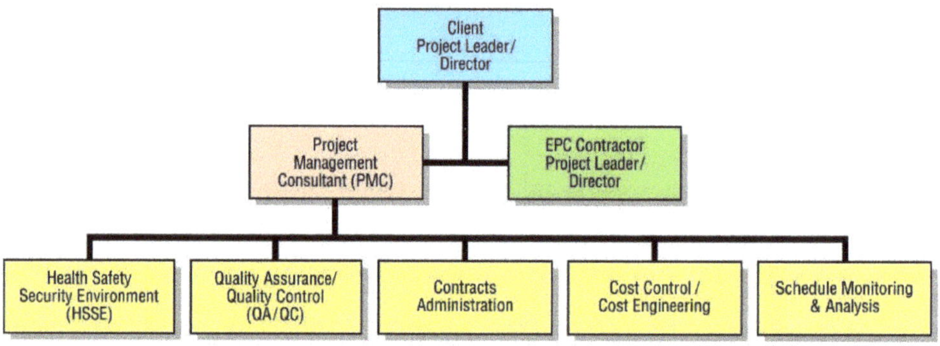

In projects in developing countries, there are several examples of the owner ceding total responsibility to a PMC for successful project delivery, despite the existence of a project leader from the owner's management. In these situations, the project leader is largely a figurehead if he or she does not possess the requisite experience to effectively drive a major multibillion-dollar process plant project. This situation is fraught with thorny issues as the project outcome in critical areas such as equipment selection, QA/QC, HSSE, project handover documentation, training, commissioning/start-up, and project closeout are at the mercy of the PMC's commitment to the highest standards of project management.

A typical structure is shown in Figure 13-3 below.

Figure 13-3: Project Organization Leadership Structure: PMC Performs Project Management and Directs EPC Contractor

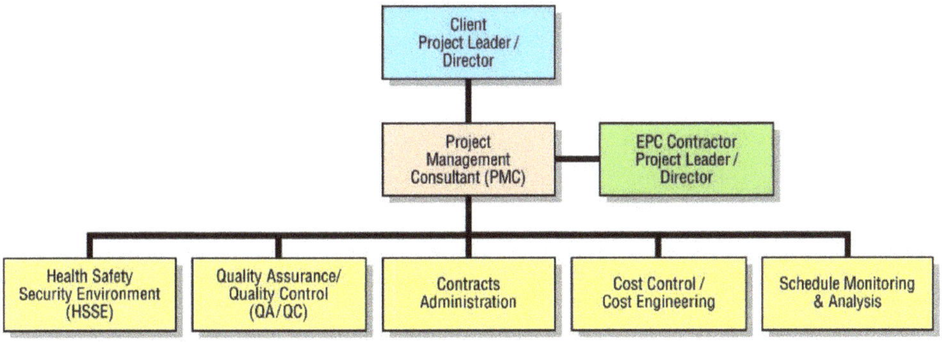

It has been the author's experience that this is not unusual in projects executed in countries with emerging hydrocarbon economies and, in every instance, there has been a deleterious impact on post-handover operations. There have also been instances of projects failing to achieve meaningful transfer of technology to local resources, particularly at the engineering level. In other words, they fail to deliver on local content provisions (or owner expectations).

13.2.3 Establish the Process by Which the Project Life Cycle Will Be Managed

The owner's project leader/director is responsible for selection of the project life cycle management process, which is typically approved by the owner's project oversight committee or the project sponsor. It is generally accepted as best practice that the stage-gate approach, once adopted with the required rigor, will deliver a successful project.

In a Stage-Gate Process, there are defined activities that must be executed in each stage, and the project is only authorized to progress from one stage to the next with the approval of a "Gatekeeper," who could be an individual or a committee. The selection of the gatekeepers for each stage is a critical success factor for a Stage-Gate Process to deliver a successful project.

Why a Stage Gate Process? There are many and varied reasons, but in the author's experience, this type of project management approach is greatly beneficial because a Stage Gate Process:

- Is a proven approach;
- Achieves cost and schedule improvements of ten percent and potentially more per specific industry benchmarking;

- When applied effectively, minimizes the potential for delay and disruption claims;
- Is based on principles of other established quality systems, *e.g.*, PDCA (Plan-Do-Check-Act), ISO 9000, and Six Sigma;
- Provides a common language and framework that enable systematic project management over the project life cycle;
- Is applicable to any project, but large and complex projects (*i.e.*, major CAPEX expenditures, high-technology projects, brownfield projects in large petrochemical operating plants, offshore oil and gas projects, etc.) will require more sophisticated use of the stage-gate process principles;
- Ensures that all the work in one stage is fully completed to the agreed-upon standard before the project proceeds to the next stage;
- Implements a thorough front-end loading (FEL) plan, which is essential to successful project outcomes, *i.e.*, Plan the Work, Work the Plan;
- Establishes key project roles, *e.g.*, Gatekeeper for each stage, project leader/director, project leadership team;
- Defines project plan and business case;
- Manages stakeholders;
- Ensures project handover is efficient, with particular importance accorded to project handover documentation;
- Captures lessons learned; and
- Publishes a project investment analysis (PIA) report that records project achievement versus pre-agreed-upon project drivers.

While the number of stages and the associated activities in the management of a project life cycle may vary for each type and size of project, essentially there are at least five stages that should be considered:

1. Appraise
2. Select
3. Define
4. Execute
5. Operate

These stages are shown pictorially in Figure 13-4 below.

Figure 13-4: Discrete Stages of a Typical Project

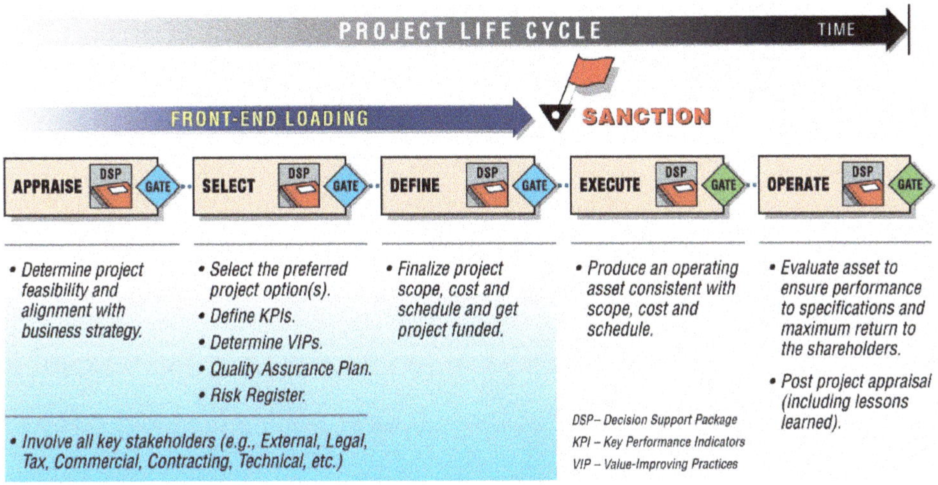

A key aspect of this type of process is to ensure that throughout the project life cycle, the appropriate stakeholders are intimately involved/consulted.

Figure 13-5 below depicts how this occurs on projects that employ project management best practices.

Figure 13-5: Integrated Involvement Throughout the Project Life Cycle

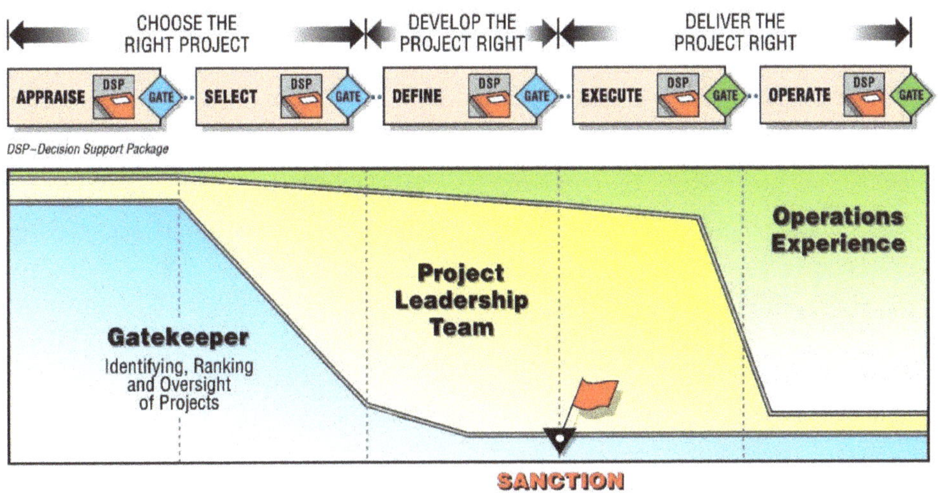

From the inception, operations personnel are imbedded in the project, which is of paramount importance for brownfield projects but also particularly important on a greenfield project. Of note is the role of the Gatekeeper, which is more critical in the early stages of the project up to project sanction, thereafter reducing but still present to the end of the project. The project team is engaged right to the end of the project, as the project handover and closeout are top priorities.

Figure 13-5 summarizes the three basic keys to successful project development and execution:

1. Choose the right project (Appraise/Select Stage)
2. Develop the project right (Define Stage)
3. Deliver the project right (Execute Stage)

By adopting and following the tenets of a Stage Gate Process, the required front-end loading (FEL) allows for efficient project development with minor expenditure up to project sanction. Figure 13-6 shows the critical importance of FEL in the early stages of the project. Conversely, the project CAPEX expenditure increases significantly after sanction when the project is in EPC or Execute Stage.

Figure 13-6: Front-End Loading Influence vs. Project Expenditures

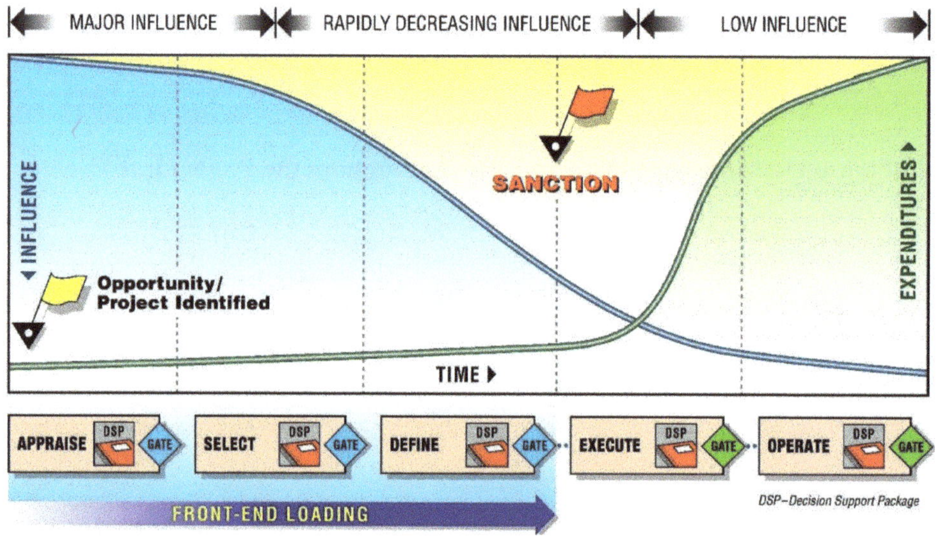

The activity sets in each of the above stages are further detailed below.

13.2.3.1 Appraisal or Concept Development Stage

At the initiation of the project, the value drivers will be established. For example, in one particular major project in a hydrocarbon-producing company, a decision was made to let the major value driver be the development of a "green" project.

In this stage, several options for the project are considered and evaluated utilizing technical and commercial criteria, and which have considered all risks that could materialize during project execution. The Appraise Stage Gatekeeper has direct control over the concepts selected for appraise stage evaluation, which should consider technical and commercially viable concepts/options with the highest standards of HSSE accorded the appropriate priority. Government authorities may require special regulatory approvals, such as a Certificate of Environmental Compliance (CEC), which in some jurisdictions can take as long as six to eight months.

The project leader/director also has the authority to request cross-disciplinary assistance from elsewhere in the owner's organization.

The Appraise Stage specifies whether the project is greenfield or brownfield. This specification is made because different considerations attend each, such as the high priority placed on interface management between the project and operations activities on a brownfield project and possibly less soil investigation required if geological mapping took place on the site during initial plant construction.

Technology selection is also included here. There could be many considerations, but if the project is brownfield, there may be a high priority on maintaining technology similar to what is in the existing plant where the new project is being contemplated. A preliminary scope of work (SOW) is prepared, and for the brownfield case, the project leader/director must ensure that a fit-for-purpose SOW is developed.

For either a brownfield or greenfield project, there must be the appropriate FEL, which should include a number of best practices, such as the following.

- Implementation of Value Improving Practices (VIP). One example of a VIP on a liquified natural gas (LNG) project would be to establish the CO_2 emissions target, usually expressed in tons CO_2/ton LNG. Generally, the lower the target, the higher the CAPEX expenditure, but a lower target may be highly valued in certain jurisdictions;
- Development of Project Assurance Plans;
- Establishment of a Preliminary Risk Register. Particular attention must be given to the management of the interface with operations;
- Field verification of as-built piping and instrument diagrams (P&IDs), one-line electrical diagrams, piping isometrics, general arrangement (GA) drawings, control system logic, alarm management, etc.;
- Integration of maintenance turnaround (TAR) activities with potential project tie-ins;
- Analysis of existing instrumentation systems;
- Development of material management plans and revision of warehouse stock levels—reorder points/economic order quantities (ROPs/EOQs);

- Inspection and testing of existing equipment. Failure to so do may lead to unwelcome surprises during construction work with a resulting potential not only for cost increases but also schedule delay and major claims that are rooted in opportunity loss (*i.e.*, production loss);
- Integration of design package delivery with construction. Identify critical engineering review items, *e.g.*, HAZOP and HAZID; and
- Evaluation of existing equipment specifications to establish the extent to which similar equipment could be procured.

Whether the project is greenfield or brownfield, the project leadership should consider retaining the expert services of companies or consultants that specialize in examining the degree of FEL on a project. Typically, the mechanisms for examining a project's quality assurance plans are a set of project input/driver frameworks developed through research performed by the consultant/company and application on a wide range of projects over many years that has built up a very significant database.

The QA plans would be evaluated in the following key areas:

- Front-end loading status. This establishes the project's FEL index, which is typically shown on a scale that compares the best-in-class QA FEL index to the worst-in-class, thereby allowing project leadership to have a sense of how its FEL measures in comparison;
- The level of application of Value Improving Practices;
- Application of lessons learned from similar projects around the world. Following the startup of a major petrochemical project, there was a failure on piping that led to the plant being out of commercial operation for almost two years, leading to billions of dollars in losses. It was subsequently discovered that failure to consider a design criterion was the root cause. A comprehensive lessons-learned session prior to engineering pre-FEED, with experienced subject experts, would have identified the necessity to apply the overlooked criterion;
- The status of the project team (the team development index).
- Project team plans for project controls (the project controls index).

In addition to these project input/driver metrics, there would be an examination of how the project would manage the introduction of new technology and how its transfer to the operations team would occur. The selection of the proposed contracting strategy would also be scrutinized to identify weak points that require further review and modification.

At this stage, there is definite value-add to considering a benchmarking exercise of similar projects to provide additional or refined evaluation criteria to aid the selection of the preferred project concept or option. The output from this stage is a small number of selected concepts or options from among the shortlisted ones that best satisfy the evaluation criteria, and the evaluation process is detailed in the Appraise Stage Decision Support Package (DSP) that is forwarded to the Gatekeeper for review and comment.

The Appraise Stage DSP would contain, *inter alia*, the following:

- Numerous concepts/options for further analysis in the next stage (Select).
- A proposed Project Charter.
- A proposed project organization structure.
- A business case.
- Project drivers.
- Quality assurance/quality control plans.
- An analysis of competing technologies.
- A plan to transfer technology.
- A preliminary project execution plan (PEP), including contracting strategy.
- Identification of key proactive measures to minimize contract disputes, including inputs from experienced claims prevention specialists.
- Proforma contracts for various works intended in the select stage, *e.g.*, pre-FEED, VIP/lessons learned workshops.
- Class four estimates of CAPEX expenditure.
- A level one project master schedule (PMS), showing efficient integration of various activities (*e.g.*, contractor, project team, owner) prepared using standard scheduling software—*as utilized on international projects*—that facilitates effective monitoring of project progress and project claims analysis arising out of contract disputes.
- A preliminary project risk register with risk mitigation strategies carefully mapped and the quality assurance plans.
- A preliminary financial model showing a base case, low case, and high case for each option, and which should be constructed to facilitate easy toggling of input parameters to instantly show their impact on the project return on investment (ROI).

- A communications plan. The project leader shall drive the completion of this plan to ensure that all stakeholders are effectively engaged. The plan typically contains important elements such as the following:
  - Engaging the fence-line community;
  - Legal and regulatory requirements;
  - Any site-specific or region-specific local governance processes such as employment policies and special procedural requirements;
  - The owner's communications policies;
  - Ensuring protection of the owner's brand, *i.e.*, reputation;
  - Plans for risk mitigation;
  - Applicable standards that will inform various aspects of project work, *e.g.*, ANSI, API, NFPA, ISO 9000, ISO 14,000, etc.;
  - An HSSE plan;
  - Procurement principles. This is especially important in developing countries where local content is of great significance to the host country; and
  - In an operating company, advance notice that the project will require human resources to support the project during its life cycle, especially to ensure coordination on a brownfield project.

The Gatekeeper may make any of the following decisions:

- Allow the project to proceed to the next stage.
- Place the project on hold.
- Kill the project.
- Recycle through the current stage, *i.e.*, send the Appraise Stage DSP back to the project leader for amendments, as he or she deems necessary to strengthen the package and business case.

Ultimately, proceeding to the next stage requires the Gatekeeper's approval.

13.2.3.2 Select Stage

In the Select Stage, the concepts/options developed and approved by the stage one (Appraise Stage) Gatekeeper are examined in far greater detail to select the single concept/option that will go past the Select Stage Gatekeeper to the Define Stage.

Typically in the Select Stage, there will be a deep dive into all the elements described in the Appraise Stage, but in addition, the project's business objectives, vision and mission, and the final selection of project drivers are detailed.

The Select Stage DSP essentially specifies the preferred concept/option to be further detailed and developed for project sanction in the Define Stage.

In the main, the activities in the Select Stage include the following:

- Finalize the Project Charter.
- Update the PEP.
- Further enhance (*i.e.*, almost one hundred percent complete) the overall project risk register and then identify the top ten risks and their mitigation strategies. Some owners wish to see this displayed as a project risk matrix, accessible to the owner online.
- Develop a detailed risk management plan that lays out the monitoring, implementation, and feedback process to ensure the identified risks are effectively managed.
- Perform a detailed (specialist) review of lessons learned from similar projects and that have to be considered in the early engineering.
- Select the final VIPs. On some projects, it is more efficient to hold a workshop to finalize the selection of drivers from lessons learned and identify which VIPs would be applied to the project life cycle.
- Select technology. Also consider how the technology is to be integrated into existing operating plants in a brownfield situation, or for greenfield projects, how the technology selection would impact requirements for training, project documentation, and handover.
- Develop an assurance register.
- If it is a brownfield project, develop the required management of change (MOC) procedure as part of the integration of the project into operations. For some years now, the technology has moved to e-MOC (software for management of change), which makes the process easier to administrate. The MOC is required to ensure that the operations group can plan effectively to take over the commissioning, start-up, and commercial operations of the new (or additional) plant, *e.g.*, review manpower levels, warehouse management, and new technology.
- Develop contracting strategy (including local content).
- Clearly define a pre-commissioning and commissioning strategy, including training and handover documentation.
- Prepare a pre-FEED invitation to bid (ITB) package, bid, evaluate, and award, and then undertake pre-FEED work to produce a basic engineering package (BEP) that informs the FEED scope to be performed in the next

stage (Define). Selection of technology for a petrochemical process plant is finalized in the pre-FEED work.

- Prepare class three (± 30 percent) CAPEX estimates.
- Update the financial model to reflect more accurate class three estimates.
- Identify long-lead procurement items; prepare purchase requisitions and obtain preliminary cost and delivery schedules from major equipment vendors. This information is fed into the class three select stage CAPEX estimates and level two PMS.
- Complete a level two PMS that shows the overall project broken down into its major components and includes the level one information expanded to show activities by area or major items of capital equipment. Very importantly, the level two schedule should show the project critical path and also the project sub-critical paths. It is necessary to utilize the services of an independent expert schedule analyst to produce a schedule report that is submitted along with the level two PMS. Such an expert provides the requisite due diligence to the identified project critical and sub-critical paths, and the logic used in the derivation of the schedule. This due diligence is a valued input into the project risk identification and management and overall project success.
- Perform Monte Carlo analysis of the class three estimate and level two PMS to ensure that the cost structure is fit for purpose and that the appropriate cost and schedule contingencies are included.
- On a brownfield project, prepare an interface management plan in consultation with the operations representatives to ensure any possible TAR activities are optimized to facilitate project tie-ins.
- Prepare an ITB for FEED scope; identify a list of qualified potential EPC contractors.
- Identify major project milestones.
- Prepare procurement packages based on expected equipment selection from the pre-FEED scope. This is a major input into the cost estimate and execution schedules.
- Prepare a preliminary work breakdown structure (WBS) to be included in the preliminary PEP.
- Prepare a preliminary project document handover plan.

At the end of the Select Stage, the Gatekeeper will not approve the project to proceed to the next Stage (Define) unless he or she is satisfied that the Select Stage work and report outcomes fully satisfy the Select Stage criteria.

13.2.3.3 Define Stage

The main focus of the Define Stage is to complete the various areas of work performed in the Select Stage and finalize all documentation required to support project sanction.

The main activity sets in this Define Stage are as follows:

- Execute the FEED scope.
- With the FEED completed, request equipment manufacturers to review any previous quotations for cost and delivery schedules. The updated quotations should be market ready such that, in the event a decision is made, post-sanction, to place orders for long-lead items, these purchase orders could be released immediately. The revised equipment cost and delivery schedules are a major input into the overall project costs and master schedule.
- Develop an ITB for the EPC works to be undertaken in the next Execute Stage. If the owner does not have a qualified stable of personnel to perform a cold eyes review, an independent reputable project management and claims consulting company with the appropriate level of EPC contract preparation and claims analysis and management experience should be retained to review the proposed ITB and the proforma EPC contract to ensure that these important documents contain the appropriate contract language/clauses needed to minimize change orders and potential delay and disruption claims. If this due diligence step is bypassed, especially on major international EPC works, there are likely to be difficulties in defending against delay and disruption claims later in the project.
- Bid the EPC works.
- Evaluate the bids and the contractors' proposed execution plans and determine the choice of EPC contractor. These results feed into the revision of the overall project costs and master schedule. Engage the independent project management and claims consulting company to provide QA/QC over the bid evaluation process and any exceptions to the owner's proforma contract, etc., that may have been detailed in the response to the ITB by the EPC contractors. Examine carefully the bidders' proposed commissioning schedules, handover documentation, project execution plans (ensure usage of a systems approach for mechanical completion), and training proposals (including provision for the owner's personnel in the contractor's home office during engineering works to assure effective transfer of technology), etc.
- Update cost estimates to class 2 (± 15 percent) and establish a control budget.

- Complete a Monte Carlo analysis of cost and schedule.
- Finalize the financial model. Ensure that all ROI-impacting parameters have been rigorously tested to ensure that the model can demonstrate the ROI hurdle rate over different sensitivities. On most major hydrocarbon projects, the minimum hurdle rate is 15 to 20 percent.
- Determine the appropriate contracting strategy and contracting plan.
- Finalize the PEP and procedures. Give particular attention to managing the interface with operations on a brownfield project.
- Identify regulatory and permitting requirements and make applications for approvals.
- Complete the level 3 project schedule. Utilize the independent reputable project management and claims consulting company to perform a quality assurance review of the EPC schedules to ensure the logic is appropriate and to determine the overall project critical path and the many sub-critical paths. In this review, the consultant also investigates the contractor's use of schedule constraints, the existence of missing activity logic, inclusion of the full scope of work, reasonableness of the critical and near-critical paths, etc. This due-diligence step is a critical success factor in ensuring that, in the event of delay and disruption claims, the project leadership is well-positioned to have the requisite analyses conducted against the as-bid cost and schedule.
- Perform preliminary HAZOP, HAZID, etc.
- Update the risk register.
- Complete the HSSE plan, identifying project metrics like lagging and leading indicators; develop control of work (COW) procedures to ensure safety of personnel and equipment during the Execute Phase.
- Update the list of major milestones.
- Prepare a preliminary pre-commissioning, commissioning, and start-up procedure.
- Complete QA/QC plans.
- Complete project controls procedures (cost monitoring and control, change order management including cost engineering, planning and scheduling, contracts administration, etc.).
- Freeze the project scope.
- Complete the e-MOC.

- Propose a document management system (DMS). Establishing such a repository of project information is critical on any project, but particularly on a large, complex project. Access to design information, construction issues, equipment evaluations, etc. are just a few examples of information that operations greatly needs post-handover. Because of the sheer volume of documentation on a large project, it may be prudent to consider a DMS that has operands (after barcoding the documents) to facilitate easy retrieval. It is a most challenging task to retrieve information from a manual system in a large room with countless filing cabinets.
- Prepare the Certificate of Environmental Compliance (CEC).
- Prepare a preliminary emergency response plan.
- Complete the PEP. This plan must contain all the elements required to assure the Select Stage Gatekeeper that the project can achieve project sanction.

At the end of the Define Stage, the Gatekeeper will not approve the project to proceed to the next stage (Execute) unless he or she is satisfied that the Define Stage work and report outcomes fully satisfy the Define Stage criteria. The end of the Define Stage represents a major project milestone, as approval to move to the execute stage effectively sanctions commencement of major CAPEX expenditures. The achievement of project sanction is a major milestone and, as shown in the earlier charts, it signals the transition of influence from the Gatekeepers to the project leader/director and the project leadership team.

13.2.3.4 Execute Stage

In this stage, the major activity is the EPC execution. Arguably, depending on the jurisdiction in which the project is being undertaken, the first order of execution is to immediately apply for a CEC (Certificate of Environmental Compliance), since the approval could require as long as six to eight months.

The activities typical to this stage are listed below, not in sequential order as many are executed concurrently. Either the EPC contractor or the owner's project leadership team are responsible for these activities.

- Award the EPC contract. Again, ensure that the independent project management and claims consulting company reviews the final EPC contract documents prior to their execution. An often-overlooked clause or requirement (or one that is properly specified) in the contract documents is project handover documents.
- Finalize the level 4 and level 5 schedules. Ensure that the EPC contractor submits an as-planned (AP) schedule in electronic format as part of its bid

documentation. The importance of this requirement cannot be overstated, since all subsequent schedule updates submitted by the contractor for the duration of the EPC works will be compared to the AP schedule.

- Ensure all HSSE metrics are monitored, and all site works are executed within the COW procedures.
- Ensure that the EPC contractor submits not only monthly written reports but also electronic copies of its updated project schedules. This allows the owner's schedule expert to interrogate them continually against the as-planned schedule to determine changes in logic, activity duration, and the original critical path, whether sub-critical paths are about to become critical path (*i.e.*, zero schedule float), whether the owner can robustly defend any contractor's claims for an extension of time and compensable delay, etc. This type of quality feedback allows the project leader/director to make important decisions based on factual information. From the start of EPC works when the entire schedule is as-planned, each monthly reporting period or window progressively shows the schedule becoming more as-built (AB) and less as-planned (AP).
- Confirm the EPC work breakdown structure (WBS). It is against this structure that all EPC payments will be made; therefore, project leadership must ensure that its team is properly resourced to provide a third eye on daily progress, or lack thereof, and the associated reasons for delays, etc. Failure to set this up with the right level of construction experience inevitably leads to challenges in defending against delay and disruption claims.
- Confirm the contractor's design of the systems handover process to mechanical completion. Very importantly, ensure that the contents of each systems turnover package are agreed upon very early in the commencement of EPC works. Agree on the process by which punch list items are corrected.
- Review the contractor's plan for pre-commissioning, commissioning, and startup.
- Perform reliability, availability, and maintainability (RAM) analysis for all rotating equipment.
- Develop a process control alarm management system, including cause and effect diagrams.
- Complete an emergency response plan: test on a periodic basis. This is especially critical on brownfield projects where hot work is performed either in or close to operating facilities.

- Confirm all required piping and other (*e.g.*, control systems, fire systems) tie-ins.
- Review the performance test procedure, including metrics for the final certificate of completion and formal project handover.
- Develop design drawings to Issued For Construction (IFC) status.
- Procure all equipment and materials.
- Initiate QA/QC procedures. Prepare and implement a third-party inspection plan if procurement is being done offshore. This is particularly important on projects where special materials of construction are being utilized and the testing requirements are onerous.
- Conduct HAZOPs, HAZIDs, constructability reviews, etc.
- Ensure HAZOP, HAZID, and constructability review outputs are properly documented, and all required changes implemented. As part of QA/QC, ensure that all items for action are closed out in the approved manner as per QA/QC procedures.
- Review construction plans, based on constructability reviews, etc.
- Prepare, bid, evaluate, and award construction ITB packages.
- Construct the project facilities.
- Ensure that overall project progress is documented via monthly reports from the EPC contractor.
- Initiate the change management process (*e.g.*, e-MOC), with particular focus on the management of the project interface with operations, if a brownfield project.
- Ensure that the transfer of technology plan proposed by the EPC contractor is being followed, particularly in respect to assignment of owner resources in the EPC contractor's home office.
- By working with owner operations, ensure that an asset reporting system is developed in alignment with the owner's operations practice, *e.g.*, equipment numbers.
- Ensure that the project controls unit under the project leader/director is properly resourced to begin work right from the start.
- Obtain regulatory approvals.
- Prepare and implement a document control system, including a manager with the relevant experience.

- As changes are made on various documents (*e.g.*, P&IDs, PFDs, piping isometrics, GA drawings, one-line electrical diagrams, etc.), ensure that all changes are redlined and maintained in a DMS that has controlled access. This is very important as at the end of the project, all verified changes must be incorporated into the final as-built documentation, which is critical to safe operation of the facilities.
- Complete operations and maintenance (O&M) manuals. The use of English is the international standard, and it should be specified. If the EPC contractor is from a non-English-speaking country, the contract documents should mandate that there be a translator versed in technical English. On an international project, this precautionary step was missed, leading to immense problems with the quality of the O&M manuals.
- Execute a training program for operations personnel. Ensure adequate provision for vendor training and utilize any opportunities for owner operations personnel to be seconded into similar plants for hands-on training. This is of particular relevance when the project is based on new technology, where no similar facility is available in the country in which the project is being undertaken.
- Ensure the contractor is approaching mechanical completion (MC) on the basis of systems approach methodology, as this is the proven way to achieving MC and commissioning and startup in the shortest time from the commencement of EPC works.
- Monitor the progress of project handover documentation. If not done in parallel with the EPC works, this activity is often rushed at the end of the project, leading to disputes over the quantity and quality of such documentation.
- Hold regular weekly meetings with the EPC contractor and other stakeholders, led by the project leader/director. There should be a base agenda for the weekly meeting comprising at a minimum: a project schedule showing progress on all items for the past week; a two or three-week look-ahead schedule that takes cognizance of workarounds to maintain the overall project master schedule; HSSE metric performance (leading and lagging indicators, control of work); change order log review; request for information (RFI) log review; WBS review (at the end of the month, the contractor's submission for pay, based on its earned value from the WBS, should not be contentious if the owner confirms the progress); subcontractor performance; operations issues; project documentation handover; O&M manual preparation progress; vendor progress on major equipment; on-site construction issues; and industrial relations issues.
- Achieve mechanical completion of each system; ensure the punch list is developed with operations input (*i.e.*, operations is part of the systems

walkdown teams). Ensure turnover package documentation is as per contract requirements.

- Achieve overall mechanical completion.
- Complete pre-commissioning. With the application of systems methodology for mechanical completion, it is usual to be able to first commence pre-commissioning on certain utilities systems.
- Complete commissioning. Ensure pre-start-up review is conducted. On most brownfield projects, the commissioning and start-up are performed by owner operations personnel.
- Start up and complete the performance test. Ensure that operations experts compile and review all performance test data to confirm that the plant is performing as per nameplate (design).
- Hand over the project to owner operations.

13.2.3.5 Operate Stage

Following a successful plant performance test, and project handover, the owner operations team assumes control of the operating facility.

However, as shown in Figure 13-5, the involvement of the project leadership team is not yet done, as in the Operate Stage, several important activities must be concluded, *viz*:

- Complete outstanding contractual work, including punch list items.
- Publish the official as-built project master schedule.
- Finalize agreement on all contract issues, *e.g.*, delay and disruption claims, change order costs.
- Close out all contracts.
- Capture key project lessons learned.
- Complete a performance analysis of project controls versus agreed metrics.
- Utilize an independent consultant to establish the final benchmark performance rating of the project.
- Ensure the document management system is effectively functioning with the correct operands, if an electronic system was implemented at the inception of the project.
- Complete as-built documentation.
- Complete vendor testing and documentation.

- Assess the management performance of the project leadership team and provide feedback to those human resources.
- Ensure transition of remaining project material in the project warehouse to the owner's warehouse system; dispose of any material deemed irrelevant to the owner's needs.
- Ensure that owner accounting incorporates all project assets into its financial system.
- Prepare and publish a PIA that compares project performance against the project performance metrics established at project sanction. This report also contains final cost information on all monies expended on the project, including all contracts, equipment purchases, and project return on investment (ROI), etc. The PIA is essentially the final project closeout report.

13.2.4 Conclusion

Large oil refinery, petrochemical, and offshore oil and gas projects are hugely cost-intensive, and the efficient development and execution of these megaprojects can often make the difference between a company going bankrupt or, at a minimum, experiencing serious negative financial in-flows due to schedule delays that lead to huge claims and increased costs.

On an LNG plant in 1995, for example, the cost efficiency may have been \$250–\$300/tpa in jurisdictions with well-established local construction and process plant operating capability. In 2015, however, in jurisdictions with new discoveries of natural gas, where there was no similar experience, the cost efficiency could be as high as \$3,000/tpa. A world-class ammonia plant requires a CAPEX expenditure of close to \$2–2.5 billion, similar to that required for a methanol plant.

It should, therefore, be patently clear that a tried and proven project management process, if properly followed as outlined above, results in a successful project with a minimum of claims and cost and schedule overruns.

This chapter provides guidelines for effective project implementation. If an owner company desires to implement a Stage Gate project management system, it should engage an independent consultant with hands-on experience setting up such a system starting from a blank sheet.

Chapter 14

Project Team Motivators and Demotivators

Chapter 14

Project Team Motivators and Demotivators[41]

The leader of a corporation or project is the individual who must ultimately be willing to take responsibility for results. Within the context of an organization or team made up of individuals, it is the collective performance of the individuals, as a team, that defines the results of the whole. While it can be said that the best motivation is internal motivation as opposed to external motivation, the leader is ultimately the one responsible for creating the conditions where motivation can thrive.

In any job, there are circumstances that motivate and those that demotivate. It is up to the leader to know which is which. According to the Pareto principle (the 80/20 rule), it is essential for the leader to know what factors most contribute to results, to help guide his or her focus.

This section discusses: 1) typical motivators and demotivators on construction sites; 2) the foreman delay survey, which is a good way to identify and eliminate obstruction factors and demotivators; 3) Herzberg's principles concerning motivators and demotivators; and 4) the author's experiences from managing construction projects and life in general.

14.1 Motivators and Demotivators

Below is a list of motivating factors from a survey of twelve construction sites. It is recommended that those who work in construction make sure to have these things in place:

- Good relationships among all people involved
- A good orientation program
- A good safety program
- Pay
- Overtime pay
- Recognition

[41] This section is a slightly modified chapter from *On Time and Under Budget: The Ultimate Power of Team Leadership*, by Michael J. Vallez, P.E., M.B.A., Lean Six Sigma, published in 2014.

- Defined goals
- An open house and a project tour
- A well-planned project
- Suggestions are solicited

The demotivators, things to avoid and/or eliminate, are as follows:

- Disrespectful treatment
- Lack of material and tool availability
- Having to do work over
- Crew discontinuity
- Project confusion
- Lack of recognition
- Ineffective utilization of skills
- Incompetent personnel
- Lack of cooperation among crafts
- Overcrowding
- Poor inspection programs
- Communication breakdowns
- Unsafe conditions
- Lack of participation in decision making

14.2 Foreman Delay Survey

In the construction industry, the foreman delay survey is a way to identify and eliminate obstructive factors and demotivators. Figure 14-1 describes the following steps.

- First, the foreman completes a daily survey of project delays for a week so as to identify what is causing the holdups.
- Second, the delays are evaluated, solutions brainstormed, and improved working methods implemented.
- Finally, the results of the new ways are analyzed, and the cycle starts again, if necessary.

Figure 14-1: Foreman Delay Survey Cycle

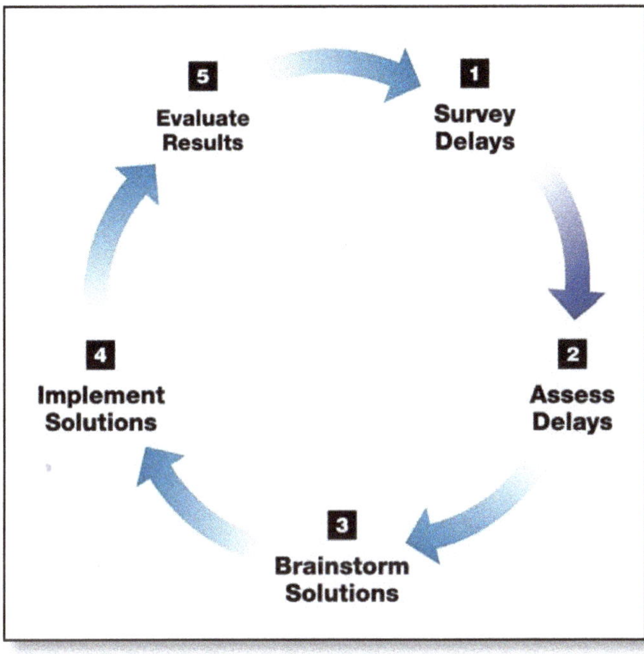

One might think that a good supervisor does not need to do this, but the author has applied these principles to projects thought to be going smoothly and has always found that the foreman delay survey makes a difference. Even the front-line foremen are surprised when they quantify their delays.

For example, on one project, a superintendent did not want to hear about problems from the workers or the foremen for some reason. People were pulling aside the project manager to report problems they were having in the field. The project manager noticed the superintendent's behavior whenever someone under his command raised a problem. He took it as a personal attack instead of a call for help. This was an ideal situation for the foreman delay survey.

Figure 14-2 shows the results from the first week of that survey. The total delay time was nine percent of the total available working hours. Frustrating delays can have a compounding effect on labor motivation. This is because as workers become more frustrated, they lose focus on safety and their jobs, leading to more frustration and more delays. In addition, frustration is one of the four mental states that lead to accidents. Most delays are management controlled, so management should use this technique as a listening tool to uncover problems. Once the problems are identified, the root causes can be identified and eliminated.

Figure 14-2: Sample Foreman Delay Survey Results

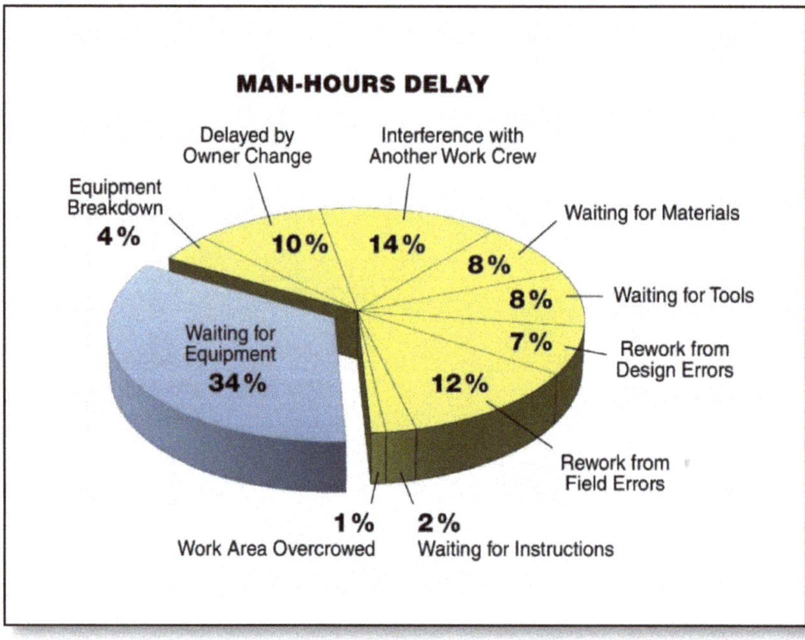

Having used this process on multiple projects, the author has found that it takes about three weeks to achieve the organizational alignment necessary to eliminate most of the delays. The process should be used frequently, if not continuously, as a project moves through various stages. Most delays are management controlled, so it is critical that management uses this technique as a listening tool to uncover the root of the problems.

When combined with goal setting and the other principles, the foreman delay survey is a simple and effective way to remove problems that lead to frustration and low morale and, consequently, a way of eliminating one of the root causes of accidents.

14.3 Herzberg's Principles

Frederick Irving Herzberg was an American psychologist who became one of the most recognized and influential names in business management. He developed what he called the Motivation-Hygiene Theory of human motivation, or the "Dual Structure Theory." According to this theory, people are influenced by two sets of factors—motivators and hygienes (demotivators), as described in Figure 14-3.

Figure 14-3: Herzberg's Motivation-Hygiene Theory of Motivation

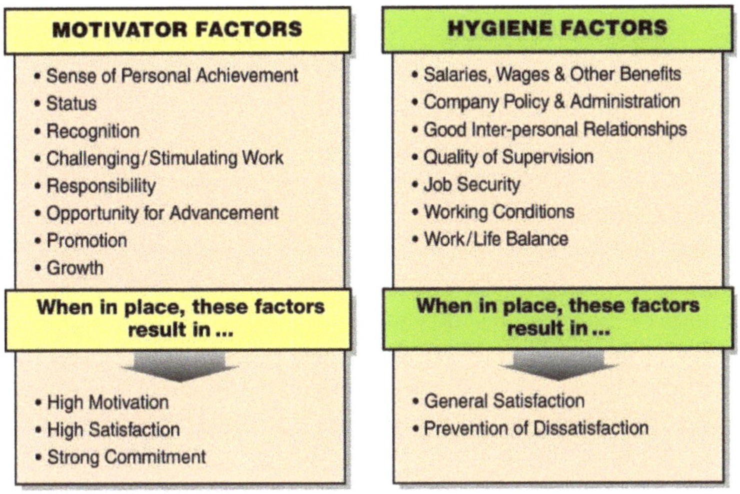

Herzberg conducted workplace surveys to identify the factors that resulted in increased or decreased motivation. The survey responses identified the factors, how frequently they appeared in the workplace, and the strength of their impact on motivation or demotivation. Not surprisingly, the factors leading to motivation are not the same as those leading to demotivation, as shown in Figure 14-3.

Another way to apply the findings of the Motivation-Hygiene Theory in practice is through vertical job loading. Figure 14-4 is a summary table of the steps that Herzberg identified for the process of vertical loading.

Figure 14-4: Herzberg's Principles of Vertical Job Loading

	PRINCIPLE	MOTIVATORS INVOLVED
1	Removing some controls while retaining accountability	Responsibility and personal achievement
2	Increasing the accountability of individuals for their work	Responsibility and recognition
3	Giving a person a complete neutral unit of work (module, division, area, etc.	Responsibility, achievement and recognition
4	Granting additional authority to employees in their activity; job freedom	Responsibility, achievement and recognition
5	Making periodic reports directly available to the workers themselves rather than to supervisors	Internal recognition
6	Introducing new and more difficult tasks not previously handled	Growth and learning
7	Assigning individual specific or specialized tasks, enabling them to become experts	Responsibility, growth and achievement

14.4 True Empowerment

Empowerment is a buzzword that is too often thrown around like a basketball without ever making it into the hoop! True empowerment means trusting people to do the right thing, and then giving them the freedom to succeed. When leadership saw how much time was being lost due to a lack of tools and supplies, it empowered employees by giving them the authority to write field requisitions for the tools and materials that they needed to be successful. One morning, after compiling the results from a Foreman Delay Survey, the project manager met with every worker on the project after the morning safety meeting and laid out the new procedure. While holding up a copy of the new material requisition book, he explained that they were now empowered to take their success into their own hands. As part of their look-ahead task planning, they were to identify the tools, materials, and equipment that they would need to perform their assigned tasks and fill out a requisition for them. The results were remarkable! There was such pent-up frustration over lack of supplies that the onrush of requests was like opening the floodgates. The requests kept the project management team busy, but it quickly saw the success of the plan as delays were virtually eliminated. Of course, before the materials or tools were purchased, they had to undergo final review and receive approval from the superintendent or project manager to avoid duplication and excess.

14.5 Life Lessons

While thinking about encouraging motivators and eliminating demotivators in the workplace, it is a good idea to look around to make sure that one is not sabotaging success by the very nature of his or her self-created environment. Whether it is a messy closet, a long overdue visit to the dentist, or other unfinished business, there are things in one's personal environment that can either empower the self or drain mental energy. These are analogous to the Herzberg motivators and hygienes that exist in the workplace. To be one's best, it is necessary to take care of these things if they need attention. One might start the process by cleaning one's closet, then mentally looking at every aspect of his or her life as a closet and doing the same. One can do what needs to be done, keep what is useful, and discard the rest.

Chapter 15

Claims Can Be Mitigated If the Owner Coordinates and Cooperates on Multi-Prime Construction Projects

Chapter 15

Claims Can Be Mitigated If the Owner Coordinates and Cooperates on Multi-Prime Construction Projects

In the construction industry, the owner makes the ultimate decision to implement a project delivery program. The project delivery program or method can take many forms. One method utilized by owners is contracting with two or more prime contractors to complete the project. The use of multi-prime contracts can have inherent problems for owners that can directly affect one or several prime contractors. In certain situations, an owner's actions or inactions may directly or indirectly impact a contractor's ability to enjoy a least-cost performance. Deficient owner coordination and cooperation is a major problem that can plague multi-prime construction projects. The purpose of this chapter is to provide awareness of the owner's duties along with case examples regarding the duty to cooperate and coordinate on multi-prime construction projects.

15.1 What Are Multi-Prime Construction Projects?

Multi-prime construction projects are simply defined as the employment of several separate prime contractors to perform distinct phases or parts of an overall project. The owner's decision to implement a multi-prime construction project is generally influenced by the following three factors, which are present in most multi-prime construction projects:

- Multi-prime construction projects are generally large in size and value and long in duration.
- The owner requires a fast-paced project in which construction starts prior to the completion of the overall design.
- The owner desires to reduce general contractor markups and fees by contracting directly with subcontractors.

In addition, multi-prime construction projects are classified into three main types. First, the owner by itself, or through a construction manager, executes separate contracts with different specialty subcontractors. Second, the owner awards separate contracts for complete phases of the project with each prime contractor responsible for all the work in its phase. Third, the owner breaks up the project site into subprojects due to the size, location, or diversity of the overall project. The owner then contracts with separate general contractors for each subproject. Example projects include the construction of military bases, penal institutions, and housing complexes.

It is axiomatic that the interdependence of the contracting parties in a multi-prime construction project is crucial to the success of the project. Coordination and cooperation, therefore, are vital to the project's successful completion.

15.2 The Duty to Coordinate and Cooperate

The duty to coordinate and cooperate is implied in every contract. This duty requires the contracting parties to deal fairly in a positive fashion and in good faith. Specifically, a contractor has the right to a least-cost performance of its contract without interference by the owner or the owner's agents. An owner's lack of coordination and cooperation can lead to schedule delay and negatively impact a contractor's least-cost performance. Typically, on a traditional design-bid-construction project, the owner and its engineer monitor the progress of the project, but they assume little or no responsibility for the coordination and performance of the work. In this form of project delivery, the contractor is typically required to coordinate and schedule its scope of work.

However, on multi-prime construction projects, the owner has an implied duty to participate actively in coordinating and sequencing the various prime contractors because the owner retains control over the performance of each individual prime contract. If a contractor can prove and demonstrate that it was delayed and disrupted due to owner-caused delay or interference caused by other prime contractors, then an owner may be liable for the delay and disruption costs. Even if the owner or owner's agent try to shift the responsibility of their duty to coordinate and cooperate through language in the contract documents, the owner may still be held liable for overall coordination of the various prime contractors. Thus, an owner must be careful and aware of its coordinating duties to ensure that its actions or inactions do not negatively impact the prime contractors' work.

A breach of an owner's implied duty to coordinate or cooperate can entitle a contractor to recover damages under court-recognized entitlements. Examples include differing site conditions, implied warranty, and maladministration entitlements. As required for all disputes, a contractor's claim for a breach of the owner's duty to coordinate or cooperate must be supported by facts. Consistent with the requirements of the contract, it is imperative that a contractor provide the owner detailed discussion and explanation including timely notice, demonstrate causation due to the owner's actions or inactions supported by contemporaneous project records, and provide damage calculations with sufficient backup documentation. The following case examples discuss how contractors were able to recover from breach of an owner's duty to coordinate or cooperate.

15.3 Case Example No. 1—Lack of Owner Coordination

In *Pierce Associates, Inc.*, GSBCA No. 4163, ¶ 12,746 (1977), the Board of Contract Appeals (BCA) found that Pierce Associates, Inc. was entitled to an equitable adjustment under the Suspension of Work clause within the contract for delays caused by another prime contractor's late performance because the government failed to demand that a slow-performing predecessor contractor comply with the government's overall schedule.

The government awarded Pierce with a Phase I mechanical contract for work on the Washington Technical Institute (WTI) facility in Washington, D.C. The government hired a construction manager and utilized a phased construction method of project delivery for Phase I of WTI. Pierce's work depended on the completion of work by the structural contractor, Piracci Corporation. Piracci's progress experienced problems that delayed Pierce.

Pierce tried to mitigate the impact caused by Piracci's delays by working around Piracci in an attempt to meet its contractual milestone dates. Pierce suffered damages to its least-cost performance due to Piracci's interference. Pierce was able to recover its increased cost under the suspension of work clause, which stated the following:

> *(b) If the performance of all or any part of the work is, for an unreasonable period of time suspended, delayed or interrupted by an act of the Contracting Officer in the administration of this contract, or by his failure to act within the time is specified in this contract (or if no time is specified in within a reasonable time), an adjustment shall be made for any increase in the cost of performance of this contract (excluding profit) necessarily*

The BCA agreed with Pierce that its contract duration was delayed by an unreasonable amount due to the government's lack of coordination. The BCA also concluded that the government failed to exercise its right of control over Piracci's inefficient work.

15.4 Case Example No. 2—Owner's Duty to Cooperate

In *Jacobson & Company, Inc.*, GSBCA No. 5605 ¶ 14,521 (1980), the BCA found that Jacobson & Company, Inc. was entitled to an adjustment to its contract due to the government's failure to prevent interferences that affected the contractor's work. The government contracted with Jacobson to perform ceiling work on a phased design and construction of a federal building in Jackson, Mississippi. Jacobson's work needed to be coordinated with a high-partition contractor that the government contracted separately.

During the project, the high-partition contractor performed its work out of sequence relative to Jacobson. Jacobson followed the sequence per the direction of the government's construction manager. This was confirmed by the government's approval of Jacobson's shop drawings that indicated the sequence. Jacobson notified the government of the following problem:

> *The braces are interfering and preventing installation of the grid and the top cap is not clear of the grid. Additionally the pre-installation of this partition is interfering with our leveling and flow of work.*
>
> *We therefore request that the braces and interfering top cap be removed and reinstalled in sequence after our ceiling is completed. Furthermore, we request your change order to compensate us for lost time…*

The government's position was that cooperation and coordination were the responsibility of each prime contractor per contract and not the responsibility of the government. Consequently, the government informed Jacobson that it was required to coordinate its work with the work of other contractors as stated in the contract documents. In contrast, the BCA found that the government failed in its duty to prevent interference with the work of Jacobson. The BCA concluded:

> *The Government can decide on whatever method of construction and contracting it chooses, but in doing so, it also assumes the responsibilities inherent in its choice. With phased construction, it is obvious that the contracting officer has the obligation and duty to demand that the various contractors cooperate with the construction schedule and not interfere with the work of any other contractor.*
>
> *It is plain that the Government is obligated to prevent interference with orderly and reasonable progress of a contractor's work by other contractors over whom the Government has control.*

Based on the BCA's ruling, Jacobson was entitled to recover increased costs from the government due to the interferences caused by the high-partition contractor.

15.5 Case Example No. 3—Unreasonable Interference by the Owner

In *United States Steel Corporation v. Missouri Pacific Railroad Company*, No. 80-1935 (1982), the United States Court of Appeals (USCA) found active interference on the part of Missouri Pacific Railroad (Mopac), which entitled United States Steel Corporation's American Bridge Division (ABD) to recover damages caused by the active interference. Mopac contracted with ABD for alterations to the superstructures of two bridges along the Arkansas River near Little Rock, Arkansas.

Since the project was a fast-track plan, Mopac contracted the substructure work in a separate prime contract with Al Johnson Construction Co. Elements of Johnson's work were dependent on ABD's work, and ABD's work depended on Johnson's timely completion of certain substructure work. Johnson's work encountered unforeseen obstacles that caused repeated revisions and extensions to Johnson's scheduled completion date. These unforeseen obstacles that impacted Johnson also delayed access and startup of ABD's work. ABD had made numerous commitments based on the timing of Mopac's notice to proceed order. Because ABD was unable to follow through with its large commitments, ABD claimed that it was damaged due to delayed access to the site.

The USCA found that Mopac, by giving notice to proceed to ABD when Mopac knew that delay to Johnson's work was inevitable, actively interfered with ABD's ability to maintain a flexible work schedule in the face of Johnson's difficulties. The USCA declared that to find active interference, the following must be demonstrated:

> *There must be findings that the party committing the interference willfully acted in bad faith which unreasonably interfered with the contractor's compliance with the contract.*

The USCA concluded that Mopac's actions constituted a willful act of bad faith and impacted ABD's least-cost performance of its work. USCA ordered Mopac to reimburse ABD.

15.6 Case Example No. 4—Owner's Failure to Coordinate

In *Stephenson Associates, Inc.*, GSBCA 6573, 86-3 BCA ¶ 19,071 (1986), the BCA found that the government was responsible for damages suffered to one prime contractor that were caused by another prime contractor, despite a clause disclaiming government liability for any costs incurred by the contractor as a result of another contractor's failure to coordinate. The claim arose out of a contract for the phased construction of the Federal Law Enforcement Training Center (FLETC) in Brunswick, Georgia. Stephenson Associates' contract was one of six prime contracts let by the government for work on the FLETC.

Under the terms of its contract with the government, Stephenson Associates was responsible for coordinating with other prime contractors. The government attempted to disclaim its responsibilities for consequences suffered by one prime contractor as a result of another prime contractor's failure to coordinate based on the following contract language:

> *The Government shall not...be liable for any costs incurred by the Contractor by reason of (an)other contractor's failure to comply with directives of the Construction Manager or the Contracting Offices, it being understood and agreed that the Government does not guarantee that other contractors will not breach their obligations to coordinate their work with that of the Contractor.*

The BCA held, however, that the government's exculpatory language did not relieve the government's responsibility to coordinate the performance and timely completion of various prime contractors. The BCA stated:

> *However, we have held that even affirmatively exculpatory language such as this does not overcome the Government's liability under the Suspension of Work clause for unreasonable delays caused by the Government's failure to coordinate the work of its prime contractors…*

Even with this exculpatory language in the contract, the government had a duty to coordinate. Further, the BCA referred to an earlier ruling from *Pierce Associates, Inc.*, GSBCA 4163, ¶ 13,078 (1978), stating the following:

> *The Government may not give such exculpatory language precedence over the Suspension of Work clause which has been mandated by the Federal Procurement Regulations. Under the terms of its contract, the Government had the responsibility to make all reasonable efforts to keep its other contractors on schedule when the timely performance of one contractor was dependent on that of others. To allow the Government to escape the responsibility imposed upon it by terms of the contract and to deny the Appellant the adjustment cognizable under a relief-granting clause would remove much of the Government's incentive to properly supervise the progress of a phased construction project.*

The BCA found that the government breached its duty to coordinate and ruled in favor of Stephenson Associates.

15.7 Conclusion

If the owner or owner's agents do not practice and provide effective contract administrative control with respect to coordination and cooperation on multi-prime construction projects, then a chaotic work environment can result. Since timely start and completion of specific work packages by one or many prime contractors may dictate the timely start and completion of subsequent prime contractors, delays and disruption issues attributed to poor owner coordination and cooperation can negatively impact the least-cost performance of the successor prime contractors. The resulting domino effect can trigger protracted claims and disputes since the owner has the duty to coordinate and cooperate.

Chapter 16

Commercial Awareness Training for EPC Projects

Chapter 16

Commercial Awareness Training for EPC Projects

The fundamental goals of the project execution team (PET) in the performance and handover of a large-scale, multidimensional EPC project include delivering the project to the owner on schedule and within budget, achieving all performance guarantees and technical requirements, and maintaining a profitable gross margin. To successfully achieve these goals, the PET must have a clear understanding of the EPC contract and the parties' obligations and responsibilities to the EPC contract. The members of the PET must also have a clear understanding of their individual roles and responsibilities in the execution of the EPC contract.

The key commercial and technical aspects of the EPC contract must be analyzed and summarized in a detailed presentation to be communicated to the PET through training sessions to instill a sense of commercial awareness regarding the EPC contract and to develop team building amongst PET members.

16.1 Contract Handover Meetings

Key members of the PET are usually different than the sales team individuals who negotiated and sold the EPC contract. It is essential that, following execution of the EPC contract and prior to Notice to Proceed (NTP), contract handover meetings be conducted between sales team personnel and at least the project director and/or the project manager as well as other key PET members (if already assigned to the project).

The purpose of these meetings is for the sales team to explain how the final EPC contract documents were developed and negotiated; to advise the PET regarding potential areas of risk in the EPC contract related to scope definition, budget, schedule, execution strategy, performance guarantees, bonuses, liquidated damages, etc.; and to suggest how to mitigate such risks in the context of the final negotiations with the owner. During EPC contract negotiations, the respective parties in the negotiations may have had certain commercial or technical issues on which they did not finally agree.

However, in the interest of "getting it done," unresolved issues may find their way into the EPC contract with both negotiating parties believing that the "creative language" included in the EPC contract resolved the issue in their respective favor. If contract handover meetings between the sales team and the PET do not take place to reconcile differences in the tendering baseline documents and the EPC baseline contract documents and identify issues that have intentionally been left open to interpretation in the final EPC contract, the PET may discover open risk issues too late to properly mitigate them and wind up with obligations that cannot be performed without additional cost and/or schedule impacts. Many project managers and EPC contract managers at some point in their careers find themselves faced with this problem. It should be noted that another benefit of proper handover meetings between the sales team and the PET is to build a trusting and working relationship between these two teams with the common goal of successfully executing the project and maximizing gross margin.

16.2 Building Blocks of Commercial Awareness Training

Prior to the issuance of the Notice to Proceed, the key members of the PET should be appointed. On a large-scale, multidimensional EPC construction project, this would generally include a commercial manager and EPC contract manager. The construction industry has recognized the value of EPC contract managers for direct commercial support and guidance to the project director and project manager. The EPC contract manager provides proactive administration and management of the EPC contract and commercial coordination between the PET and the owner. In this regard, the EPC contract manager has the lead responsibility to develop the commercial awareness training for the PET. If an EPC contract manager is not assigned to the project, the responsibility for commercial awareness training would normally fall to the project manager (along with all the project manager's other responsibilities). Alternatively, if no in-house expertise is available, the project could engage an external consultant to prepare and present the commercial awareness training for the project team.

The development of the commercial awareness training presentation includes the preparation of an EPC contract brief that summarizes each clause of the EPC contract general conditions, special conditions, and commercial exhibits. Once this fundamental tool has been developed, it can be used to prepare specific reference lists for the PET such as:

- Action requirements
  - Reporting
  - Submittals
  - Meetings
- Notice requirements
  - Change event that may result in schedule extension and/or additional compensation
  - Intent to subcontract or procure major equipment over a certain dollar value
  - Shipping notices
  - Potential insurance claim
  - *Force Majeure* event

  - Intent to work on public holidays
  - Start of performance tests

- Owner obligations
  - Site access
  - Permits
  - Construction all risk insurance
  - Construction water and power

- Change order procedure
  - Change identification
  - Notice
  - Documentation

16.3 Key Items of the Commercial Awareness Training Presentation

After the contract handover meetings with the sales team and the preparation of the basic reference documents (contract brief, action requirements, notice requirements, owner obligations, etc.) have been completed, the EPC contract manager should focus on the preparation of the commercial awareness training presentation. In general, the presentation should include a review of the following items:

- Where to find the EPC contract documents in the project document control system, including the tendering baseline documents and the order of precedence of these documents under the EPC contract
- Owner's organization and who on the owner's team has authority to authorize and approve changes
- Owner's obligations
- Action requirements list
- Notice requirements list
- Scope of work
- The critical path and key milestones of the project schedule
- Performance guarantees, potential bonuses and incentives, and potential liquidated damages
- Payment provisions and required supporting documentation
- Project insurance provided by the owner

- Change order procedure with an emphasis on how to recognize a change or compensable event, notice requirements, and required supporting documentation

16.4 Preparation for the Commercial Awareness Training Meeting

When the presentation has been prepared and approved by the project director and project manager, it should be distributed to the PET members for information prior to the commercial awareness training meeting. PET members should be advised to read the EPC contract, commercial exhibits, and scope of work, and become familiar with the EPC contract brief and other commercial tools such as actions requirements, notice requirements, and the change order procedure, prior to attending the commercial awareness training meeting.

The PET members should also be requested to identify contractual obligations that, based on their expertise and team function, pose potential risks to the project schedule or cost during performance of the work if not managed properly.

For example:

- The procurement manager may advise of a possible schedule risk due to the late delivery of a major equipment component.
- The engineering manager may advise of a possible technical risk due to an ambiguity in the performance test criteria that the owner could interpret in such a way that the performance test would never succeed as written.
- The subcontracts manager may advise of possible cost risks due to higher-than-budgeted tender responses from potential subcontractors.

16.5 Objectives of the Commercial Awareness Training

The objectives of commercial awareness training should be to:

- Increase the PET's knowledge of the essential commercial features of the EPC contract and, through better utilization of contract management tools, heighten the identification of entitlement to time extensions and additional costs. The commercial awareness presentation should:
    - Inform the PET where it can find all baseline EPC contract documents within the project's document control system for reference as necessary.

  - Provide the reference tools for action requirements, notice requirements, the change order procedure, etc.
  - Review key commercial and technical elements of the EPC contract such as the critical path, insurance provisions, payment provisions, performance guarantees, bonus provisions, liquidated damages, and the owner's obligations and supply scope.

- Obtain feedback from the PET regarding possible cost, schedule, and technical risks.
  - Each PET member will have the opportunity to identify potential risks to the project and initiate discussion on how such risks can be mitigated.
  - A member of the sales team that negotiated the final EPC contract should also be invited to the meeting to provide feedback on any issues that were not fully clarified or documented during final negotiations.

- Provide a team-building experience by clarifying duties and responsibilities regarding management of risks, provision of notices to the owner, follow-up of contractually defined actions and support of the change order procedure.
  - After discussing notices, actions, risks, and the change order procedure, etc., the PET, led by the project manager, should agree on which team member is responsible for carrying out actions, issuing notices, and tracking potential risks. This should be documented by the EPC contract manager, issued to the PET, and updated as necessary during the project.
  - The PET should be encouraged to proactively participate in the commercial management of the EPC contract and to communicate ideas and observations regarding risk mitigation, potential "scope creep," or any event that may affect the successful delivery of the project to the owner or impact the project's gross margin.

16.6 Conclusion

Commercial awareness training for PET members on large-scale, multidimensional EPC construction projects is an essential activity that will encourage team building and communication, identify potential project risks and mitigation strategies, establish responsibilities between the PET members, and strengthen change identification and the change order process. The development of a detailed and effective commercial awareness training presentation is an arduous and time-consuming task that the project director and project manager in many cases "do not have time for" at the beginning of a project. If there is no in-house expertise to perform commercial awareness training, the project should consider hiring a qualified consultant to conduct the training for the PET.

Commercial awareness training should be provided periodically over the life of the project as new members join the team. For example, the training should be performed in advance of NTP, at the start of construction, during construction, and at the start of commissioning and performance testing. It is also a good practice to conduct a redacted form of commercial awareness training with the major on-site civil, mechanical, and electrical subcontractors for the purposes of claims avoidance and team building.

Chapter 17

Contractor's Risk Analysis

The best way for a construction contractor to prevent disasters and claims on a project is to evaluate thoroughly, before the job is bid, the risks involved in performing the project and implement necessary and effective mitigation measures in its management of the project to control these risks.

A Risk Analysis Checklist is provided as Table 17-1 and includes the following major categories of issues:

- Precontract/Performance issues
- State lien law requirements
- Bond requirements
- Critical contract clauses
- Notice requirements
- Scheduling requirements
- Change order procedures
- Payment provisions
- Disputes procedure
- Federal government contracts

Table 17-1: Contractor's Risk Analysis Checklist (12 Pages)

LONG INTERNATIONAL

CONTRACTOR'S RISK ANALYSIS CHECKLIST

PRE-CONTRACT / PERFORMANCE ISSUES REQUIREMENT / ACTIVITY / RISK	YES / NO
City, County, or State Licensing Requirements	☐ ☐
Foreign Corporation Registration	☐ ☐
Non-Resident Contractor's Bond or Registration	☐ ☐
Annual Filings / Reports	☐ ☐
Sales / Use Tax Obligations	☐ ☐
Other Taxes	☐ ☐
County Municipal Requirements	☐ ☐
Permits / Fees	☐ ☐
Insurance Requirements	☐ ☐
Environmental Site Assessment	☐ ☐
Financial Capacity of Parties	☐ ☐
Parties' Performance History	☐ ☐
Prequalification Requirements	☐ ☐

Page 1 of 12

Table 17-1: Contractor's Risk Analysis Checklist (continued)

LONG INTERNATIONAL

CONTRACTOR'S RISK ANALYSIS CHECKLIST

STATE LIEN LAW REQUIREMENTS REQUIREMENT / ACTIVITY / RISK	YES / NO
Pre-Contract or Pre-Performance Filing Required	☐ ☐
Pre-Contract or Pre-Performance Notice Required	☐ ☐
AIA 201 & 2.1.2 Request to Owner for Property Description	☐ ☐
Obtaining Legal Description	☐ ☐
Preliminary Notices	☐ ☐
Deadline for Filing Lien	☐ ☐
Definition: "Last Labor or Materials"	☐ ☐
Scope of Lien Law Coverage	☐ ☐
Deadline for Lien Foreclosure	☐ ☐
Miscellaneous Notice-Filing Requirements	☐ ☐
Contractual Lien Waiver Clauses	☐ ☐
Pay Application Lien Waivers	☐ ☐
Property Exempt from Liens	☐ ☐
Lien Subordination Agreement	☐ ☐
Bonding off Liens	☐ ☐
Final Payment Lien Waivers	☐ ☐

Page 2 of 12

Table 17-1: Contractor's Risk Analysis Checklist (continued)

LONG INTERNATIONAL

CONTRACTOR'S RISK ANALYSIS CHECKLIST

BOND REQUIREMENTS REQUIREMENT / ACTIVITY / RISK	YES / NO
Bid Bond Required	☐ ☐
Performance Bond Required	☐ ☐
Payment Bond Required	☐ ☐
Bond Notice Requirements	☐ ☐
Contractual Time Limit for Suit on Bond	☐ ☐
Statutes Affecting Bond Claims	☐ ☐
Comparison of Bond With Statutes	☐ ☐
Is Contract Incorporated Into Bond?	☐ ☐
Scope of Bond Protection	☐ ☐
Arbitration With Principal and / or Surety	☐ ☐

Page 3 of 12

Table 17-1: Contractor's Risk Analysis Checklist (continued)

LONG INTERNATIONAL

CONTRACTOR'S RISK ANALYSIS CHECKLIST

CRITICAL CONTRACT CLAUSES REQUIREMENT / ACTIVITY / RISK	YES / NO
No-Damages-for-Delay / Inefficiencies	☐ ☐
Changes Clause	☐ ☐
Differing Site Conditions	☐ ☐
Site Investigation Obligations	☐ ☐
No Claims Due to Site Conditions	☐ ☐
Pricing of Significant Quantity Variations	☐ ☐
Economic Price Adjustments	☐ ☐
Shift of Design Responsibility	☐ ☐
Drawing Review Obligations	☐ ☐
Shop Drawing / Submittals Procedures	☐ ☐
Suspension of Work Rights	☐ ☐
Termination for Convenience	☐ ☐
Termination for Default	☐ ☐
Overhead / Profit Limitations	☐ ☐
Pay-When-Paid-Provisions	☐ ☐
Coordination Responsibilities (Design / Schedule)	☐ ☐
Warranty Risks	☐ ☐

Page 4 of 12

Table 17-1: Contractor's Risk Analysis Checklist (continued)

LONG INTERNATIONAL

CONTRACTOR'S RISK ANALYSIS CHECKLIST

CRITICAL CONTRACT CLAUSES *(continued)* REQUIREMENT / ACTIVITY / RISK	YES / NO
Notice Requirements	☐ ☐
Unusual Indemnity Requirements	☐ ☐
Lien Waiver Provisions	☐ ☐
Right to Stop Work	☐ ☐
Contractor Quality Control	☐ ☐
Unusual Insurance Requirements	☐ ☐
Scheduling (CPM) Requirements	☐ ☐
Acceleration of Work Rights	☐ ☐
Liquidated Damages Clause	☐ ☐
Clauses Incorporated by Reference	☐ ☐
Documents Incorporated by Reference	☐ ☐
Order of Precedence Clause	☐ ☐
Limitations on Authority Provisions	☐ ☐
Contract Choice of Law Provision	☐ ☐
Inspection / Testing Requirements	☐ ☐
Acceptance Procedures	☐ ☐
Final Payment Provisions	☐ ☐
Disputes Procedure	☐ ☐

Page 5 of 12

Table 17-1: Contractor's Risk Analysis Checklist (continued)

LONG INTERNATIONAL

CONTRACTOR'S RISK ANALYSIS CHECKLIST

CRITICAL CONTRACT CLAUSES *(continued)* REQUIREMENT / ACTIVITY / RISK	YES / NO
Requirements for Release of Retainage	☐ ☐
Merger Clause	☐ ☐
Exclusion of Qualifications / Clarifications	☐ ☐
Payment Application Requirements	☐ ☐
Contract Milestone Requirements	☐ ☐
Interest on Late Payments	☐ ☐
Change Clause	☐ ☐
Differing Site Conditions Clause	☐ ☐
Insurance and Bonds Clause	☐ ☐
Offset Payment Clause	☐ ☐
Indemnification Clause	☐ ☐
No Damages for Delay Clause	☐ ☐
Hazardous Materials Clause	☐ ☐
Site Cleanup Clause	☐ ☐
Lien Waiver Clause	☐ ☐

Page 6 of 12

Table 17-1: Contractor's Risk Analysis Checklist (continued)

LONG INTERNATIONAL

CONTRACTOR'S RISK ANALYSIS CHECKLIST

NOTICE REQUIREMENTS REQUIREMENT / ACTIVITY / RISK	YES / NO
Time Extensions & Excusable Delays	☐ ☐
Differing Site (Changed) Conditions	☐ ☐
Changes & Extra Work	☐ ☐
Notice of Acceleration	☐ ☐
Damage to Person or Property	☐ ☐
Notice of Claims or Backcharges to Subcontractors	☐ ☐
Termination Notices – Right to Cure	☐ ☐
Insurance Claim Notices	☐ ☐
Warranty Claims	☐ ☐
Arbitration Demand	☐ ☐
Submission to Architect / CM	☐ ☐
Errors or Omissions	☐ ☐
Weather	☐ ☐

Page 7 of 12

Table 17-1: Contractor's Risk Analysis Checklist (continued)

LONG INTERNATIONAL

CONTRACTOR'S RISK ANALYSIS CHECKLIST

SCHEDULING REQUIREMENTS REQUIREMENT / ACTIVITY / RISK	YES / NO
Presence of Contract Schedule	☐ ☐
Form of Schedule	☐ ☐
Scheduling Responsibility	☐ ☐
Right to Alter Schedule	☐ ☐
Start Date (NTP) Trigger	☐ ☐
Update Requirements	☐ ☐
Technique Required	☐ ☐
Who Owns the Float?	☐ ☐
Damages for Delay (Liquidated – Actual or Both)	☐ ☐
Excusable Delays Defined	☐ ☐
Time Extension Justification Process	☐ ☐
Milestone Dates	☐ ☐
Phased Completion – Acceptance / Commencement of Warranties	☐ ☐
Completion Date – Substantial / Actual	☐ ☐
Coordination of Work Provisions	☐ ☐

Page 8 of 12

Table 17-1: Contractor's Risk Analysis Checklist (continued)

LONG INTERNATIONAL

CONTRACTOR'S RISK ANALYSIS CHECKLIST

CHANGE ORDER PROCEDURES REQUIREMENT / ACTIVITY / RISK	YES / NO
Authority to Issue Binding Changes	☐ ☐
Required Approvals	☐ ☐
Writing Requirements	☐ ☐
Constructive Changes Procedure	☐ ☐
Financing Institution Approvals	☐ ☐
Impact / Delay Cost Recovery	☐ ☐
Overhead / Fee Limits on Added Work	☐ ☐
Limits on Number of Allowed Markups	☐ ☐
Unit Prices – Quantity Variation Pricing	☐ ☐
Force Account Work Procedures	☐ ☐
Equipment Pricing Procedures	☐ ☐
Overhead / Fee on Deductive Changes	☐ ☐
Limits on Acceleration Payments	☐ ☐
Time Limits for Proposal Submission	☐ ☐
Stated Effect of Untimely Submission	☐ ☐
Audit-Subsequent Modification Rights	☐ ☐
Funding for Changes	☐ ☐
Confirming Change Directives	☐ ☐
Tracking Direct and Indirect Costs	☐ ☐
Partial Payment for Disputed Changes	☐ ☐

Page 9 of 12

Table 17-1: Contractor's Risk Analysis Checklist (continued)

LONG INTERNATIONAL

CONTRACTOR'S RISK ANALYSIS CHECKLIST

PAYMENT PROVISIONS REQUIREMENT / ACTIVITY / RISK	YES / NO
Pay Request Form	☐ ☐
Schedule of Values Required	☐ ☐
Project Schedule Required	☐ ☐
Payment for Stored Materials	☐ ☐
Timing of Payment Application	☐ ☐
Timing of Payments (Partial & Final)	☐ ☐
Pre-Conditions on Obligation to Pay	☐ ☐
Lien Waivers – Contractor's Affidavit	☐ ☐
Deadline for Payment	☐ ☐
Right to Stop Work for Non-payment	☐ ☐
Required Notifications	☐ ☐
Withholding or Backcharge Rights	☐ ☐
Effect of Acceptance	☐ ☐
Retention Requirements	☐ ☐
Reduction of Retention	☐ ☐
Pre-Conditions to Final Payment	☐ ☐
Interest on Late Payments	☐ ☐
Effect of Final Payment on Rights	☐ ☐
Approvals Required	☐ ☐
Certification of Payment to Subcontractors	☐ ☐

Page 10 of 12

Table 17-1: Contractor's Risk Analysis Checklist (continued)

LONG INTERNATIONAL

CONTRACTOR'S RISK ANALYSIS CHECKLIST

DISPUTES PROCEDURE REQUIREMENT / ACTIVITY / RISK	YES / NO
Mandatory Mediation	☐ ☐
Arbitration	☐ ☐
Litigation	☐ ☐
Venue Selection	☐ ☐
Time Limits for Making Claims	☐ ☐
Right to Join Additional Parties	☐ ☐
Alternative Disputes Procedures	☐ ☐
Incorporated by Reference	☐ ☐
Pass-Through Claim Provisions	☐ ☐
Special Disputes Procedures	☐ ☐
Limitations on Liability	☐ ☐
Interpretation of Contract Clauses	☐ ☐
Partnering Obligations	☐ ☐

Page 11 of 12

Table 17-1: Contractor's Risk Analysis Checklist (continued)

LONG INTERNATIONAL

CONTRACTOR'S RISK ANALYSIS CHECKLIST

FEDERAL GOVERNMENT CONTRACTS REQUIREMENT / ACTIVITY / RISK	YES / NO
Responsiveness of Bid	☐ ☐
Responsibility Determinations	☐ ☐
Bid Protest Procedures	☐ ☐
Bid Mistake Relief	☐ ☐
Social & Economic Programs	☐ ☐
Buy American Restrictions	☐ ☐
Davis Bacon Requirements	☐ ☐
Cost & Pricing Data Submissions	☐ ☐
Application of Cost Principles	☐ ☐
Disputes Clause Certification	☐ ☐
Other Certifications and Liabilities	☐ ☐
False Claims/False Statements	☐ ☐
Prompt Payment Act Obligations	☐ ☐

Page 12 of 12

Chapter 18

Owner and Contractor Guidelines

Chapter 18

Owner and Contractor Guidelines

18.1 Owner Guidelines

Unquestionably, the best way to handle a dispute is to prevent it altogether. The next best way is to equitably settle a claim or change order request when it arises. The parties involved in any construction project frequently have conflicting interests; however, they have one common goal: completion of the project.

The following is a list of recommendations to owners for avoiding disputes:

- Do not rush design. Give the engineering firm and design professionals adequate time to complete, check and coordinate the design, drawings, and specifications. If the contractor performs the checking or coordination, the cost will invariably be greater.
- Do not rush preparation of the plans and specifications. A complete set of plans and specifications with adequate details defining the scope of the project reduces the chance of disputes and extra work claims.
- Create a set of plans and specifications tailored and detailed to the particular project. "Off-the-shelf" specifications can create ambiguity, conflict, and claims.
- Include clearly defined change order, time extension request, and scheduling procedures in the contract documents.
- Perform sufficient subsurface exploration to inform the bidders of the conditions of the proposed work. If you do not, you will pay a premium if ground conditions are better than expected, or you will face a claim if the contractor encounters a "differing site condition." A good precontract site investigation and subsurface exploration are fundamental to the avoidance of disputes.
- The field engineer must know the requirements of the contract. The contractor will prosecute the work according to its plan. The field engineer must know what is planned as to quality, quantity, method, sequence, technique, and procedure. A knowledgeable field engineer can catch faulty work or materials before they generate failures and extra costs. A good, honest, objective inspection helps the contractor and the owner.
- Communicate. Do not make the relationship with the contractor an adversarial one. Promote a cooperative attitude and interpret the contract

equitably. Do not try to force the contractor to provide more than what is set out in the contract. However, if the contractor submits a change order or claim, the owner may need to request more information to adequately evaluate the merits of the contractor's request.

- Be timely. If a contractor requests information, design clarification, approvals of submittals or shop drawings, act with due dispatch to support the construction. Lack of timely response is a frequent source of disputes and claims.
- Require that the contractor submit the native executable schedule files for the baseline schedule, any rebaseline schedules, and all schedule updates. If the contractor submits time extension requests, require that such requests include the submission of time impact schedules in native electronic form so that the owner can evaluate the validity of such additional time requests.
- Understand the implication of denying valid extensions of time with respect to constructive acceleration and increased project costs.
- Fairly evaluate change order requests. If the project scope and specifications are ambiguous, equitably evaluate the impacts to the contractor's schedule and costs.
- Be event oriented. Address problems when they arise. Work out solutions rather than concentrating on pointing fingers. No problem is the contractor's problem alone. If it impacts the project, it is your problem, too. Try to contemporaneously resolve extra work situations.
- Assist the contractor. Do not direct the contractor. You are there to help. The contractor must build the project. You are not there to manage the work. If you do, you become responsible for any delay or disruption to the contractor's plan.
- Document the job. Document the job. Document the job. You are the eyes and ears on the project. You must contemporaneously record the job history. Without your written words, there is not opportunity to evaluate job events. Provide sufficient information of what occurs on the job every day so someone unfamiliar with the job can become fully familiar by reading your reports and diary.

Hopefully, these recommendations will help to avoid disputes, but they should also allow for a smoother project, better cooperation between engineer and contractor, and better and easier resolution of claims.

18.2 Contractor Guidelines

Contractors can also effectively manage their work to avoid or mitigate claims by following the recommendations listed below:

- Thoroughly understand the scope of work in the owner's bid package. Assign the appropriate engineering and construction disciplines to review the bid package. Submit written questions where there is ambiguity.
- Determine the level of completeness of the owner's design. Each component of the facility may not be of the same level of design detail. The contractor may have to take the designs to a further level of detail to establish a reliable basis for quantity takeoffs, equipment sizing and specification, plot plan layout, utility requirements, etc.
- Apply the appropriate design development allowances and contingencies to the bid estimate calculations. Design development allowances and estimating contingencies are two separate items. The design development allowance is a factor that is added based on the level of completeness of the design and is usually based on the contractor's experience in growth of quantities based on completing the current status of the design to final "Issued for Construction" drawings. Estimating contingencies apply to labor rates, material and equipment costs, labor productivity factors, schedule allowances for potential delays, and other potential problems or uncertainties.
- Do not count on making the job profitable only as a result of change orders that you are aware will need to be made as a result of problems with the bid package. This will quickly sour the relationship with the owner. Resolve those problems prior to contract award or include "provisional allowance for such problems" if that is the most practical way to deal with uncertainties."
- Ensure all "flowdown" clauses are included in all subcontracts. Also ensure that the subcontractors are following the same procedures for change order requests, time extension requests, notices, and other requirements that may result in additional time and costs.
- Properly evaluate all subcontractor change order and time extension requests. If the owner is not responsible for such requests, do not try to pass these along as it will damage the relationship.
- Carefully prepare an as-planned or baseline schedule. Ensure that the as-planned schedule includes the entire scope of the project, including such things as owner approvals, permitting requirements, construction of labor camps, soils investigations, punch list activities, and startup and commissioning. Ensure that the manpower loading of the schedule is

consistent with labor availability and camp sizing basis. Establish the basis for any weather delays that are built into the schedule. Ensure subcontractor buy-in to the schedule for all subcontractor work.

- If subcontractors are responsible for level 4 schedules, ensure that they are properly prepared, are free from logic and improper constraint issues, include all the required scope, and properly roll up to the level 3 schedule prepared by the contractor.
- Perform schedule quality assurance procedures on the baseline schedule to ensure that all logic is included and that constraints are not being inappropriately used to set the start or finish of activities without the proper activity sequence logic.
- Provide written timely notice for all delays and impacts per the contract requirements. Do not assume that the owner is aware of all impacts that it may have caused, or impacts caused by *force majeure* events or other parties.
- Update the project schedule for all delays and change order impacts. Record the basis for delays to the start of activities and extensions of the durations of activities. If new activities are added, ensure that such additions are not the cause of schedule delay if the reason for adding additional activities is to expand the detail of existing activities but not the addition of changed scope of work. If the reason for adding activities is changed or added work, provide clear communication of the basis for adding activities to the owner when transmitting the updated schedule. Ensure proper logic is used for all added activities and perform quality assurance reviews of all schedule updates.
- Use agreed-upon time impact analysis procedures for determining the basis for any time extension requests. Check the contract for such procedures and the agreed-upon basis for extensions of time. If none are stated, utilize accepted industry practices.
- Properly consider time-related costs in change order requests. If extended overhead costs are being requested for delays, ensure that the basis for such costs follows the contractual procedures, and that the schedule analysis considers the contractor's concurrent delays to critical path activities.
- If your legitimate time extension requests are denied, provide notice of constructive acceleration and your intent to recover your acceleration costs. Document all acceleration costs to preserve the record for your later claim.

- Provide notice for deliberate pacing delays. If there is an excusable delay to other activities and the contractor decides to pace its other work rather than "hurry up and wait," notify the owner in writing of the intent to pace its work.

- Implement procedures to obtain the necessary data to evaluate craft labor productivity. These procedures should include the frequent tracking of installed quantities and the required labor to install the quantities, by area, and by craft.

- Perform frequent productivity reviews. If productivity is not as good as was planned, determine the reasons and implement productivity improvement plans as soon as possible. Check the bid basis for labor productivity and determine if the conditions under which the work is actually being performed is different from the bid basis for productivity.

- Evaluate the cumulative impact of change orders on labor productivity. This evaluation typically requires the recording of change order man-hours compared to planned or bid basis man-hours. If the growth in man-hours is greater than 8 to 10 percent of the bid man-hours, labor productivity may be negatively affected as a result of the cumulative impact of multiple change orders. Include such labor productivity impacts in your change order pricing rather than waiting until the end of the project. If the owner will not accept such costs, prepare separate notice letters of your intent to claim such costs at the end of the project to preserve your rights for such claims.

- Communicate. Do not make the relationship with the owner an adversarial one. Promote a cooperative attitude and interpret the contract equitably. Do not try to force the owner to pay for work that is clearly included in your contract scope. Provide all appropriate information in your change order and time extension requests to adequately evaluate the merits of the change orders.

- Document all problems contemporaneously. If email is the primary mode of both internal and external communications, provide guidance to your staff as to the proper wording of such communications to ensure the facts are properly stated.

- Prepare and implement an effective document management system. Such systems are common for all projects, but they are not all adequately implemented for later efficient retrieval of information. If PDF files are being used, agree with the owner that all PDF files must be searchable. Images of communications, such as letters, meeting minutes, RFIs, progress reports, etc., that cannot be electronically searched become problematic to both parties after thousands of such documents populate the project database. It can become very expensive to later have to deal with this problem if a major dispute occurs. Access to and the ability to cost effectively review all project information can facilitate a timely and equitable resolution of disputes. Include subject files in the project database for problems and delays to your work.

Chapter 19

Construction Claims Analysis Checklist

Chapter 19

Construction Claims Analysis Checklist

The preparation and analysis of claims on EPC projects can be a complex undertaking. Multiple engineering, construction, project management, and construction management experts and legal professionals may be required either to prepare or defend against claims, depending on the nature of the issues that comprise the claim. The types of claims and counterclaims that may arise during EPC projects are diverse and can include, but are not limited to, one or more of the following:

- Delay claims, both by the contractor for extended overhead costs and by the owner for liquidated or actual damages
- Disruption/Loss of productivity claims
- Acceleration claims
- Differing site conditions claims
- Changes in scope claims, including cardinal change claims
- Constructive changes claims
- Defective and deficient contract documents claims
- Termination/Suspension claims
- Variation in quantities claims
- Gross negligence/willful misconduct claims
- Lost profits claims
- Misrepresentation claims
- Defective work claims
- Failure to perform to a standard of care claims
- Bad weather claims
- Impossibility of performance claims
- Unjust enrichment claims
- Superior knowledge claims
- Breaches of implied warranties claims

This section sets forth the basic information and tasks in performing a claims analysis, whether one is preparing a claim or defending against one. The scope, tasks, and level of detail of the analysis, however, depend on the settlement objectives (change order negotiation, negotiated settlement of a formal claim, mediation, or expert report for arbitration or litigation), as well as the monetary risk involved, *i.e.*, the cost of the claims analysis and preparation/defense must be cost effective compared to the value of the claim. Therefore, not all tasks that are listed below may be appropriate, justified, or necessary. Although this list is extensive, it is not meant to be all-inclusive, and depending on the specific claim issues and the schedule and quantum analyses utilized, other tasks may be appropriate.

19.1 Identify and Analyze Documentation for Claim Issues

1. Determine if the project documentation is reasonably organized and if a document database will be required. If the documentation is not well organized or too voluminous to review efficiently, a document database with optical character recognition (OCR) text recognition, keyword, issue, author, recipient, date, and other search fields may be necessary.
2. Obtain and review the contract terms and conditions, change management procedure, cost and schedule control procedures, and other contract documents that may affect the analysis of the claims.
3. Identify the law that governs the contract. Seek legal guidance for any contractual issues that will govern the analysis and dispute resolution based on the governing law.
4. Obtain and review the owner's request for proposal, the contractor's proposal, and any documentation related to the development and understanding of the contract terms and the contractual scope of work.
5. Identify daily reports, diaries, correspondence, emails, and other contemporaneous communications that may contain information regarding the claim issues.
6. Review any documents setting forth the contractor's claim(s) and the owner's counterclaims. Identify and describe the nature of the claims and problems incurred or alleged.
7. Interview project personnel to understand major problems encountered and the relevant documentation related to those problems.
8. Prepare issue files that contain information related to each problem and claim issue.
9. Prepare a chronological history of events related to each claim and problem.
10. Identify work areas of the project or parts of the facility affected by the claims and problems.
11. Identify change orders (approved, disputed, or pending) that the contractor has submitted. Identify notice letters related to change orders or other alleged impacting events.

12. Obtain and review the RFI log and RFI files if prepared during the project, or logs and documents related to other project mechanisms used to contemporaneously record potential impacts, such as Design Change Notices (DCNs).
13. Identify any actions or problems caused by the contractor or its subcontractors, vendors, and suppliers that may have contributed to delays, man-hour overruns, and cost overruns incurred by the contractor.
14. Identify if construction means and methods changed because of the claims and problems.
15. Identify any hindrances or interferences to the contractor's work that were caused by actions/inactions of the owner or its agent(s).
16. Identify any owner-furnished information, equipment, materials, or other items that were delivered late or were defective that may have caused delays.
17. Identify any design or specification problems (errors, ambiguities, etc.) that may have caused or contributed to the claims and problems.
18. Perform a technical assessment of any engineering, specification, construction means and methods, or other technical issues to determine the contractor's entitlement to delay and additional costs.
19. Perform an assessment of the project and construction management performance of the owner and contractor. Use a standard of care, such as that identified in the contract, or industry standards such as the Project Management Institute (PMI), Construction Industry Institute (CII), AACE International, American Petroleum Institute (API), Associated General Contractors (AGC), Construction Management Association of America (CMAA), or others to determine responsibility for any mismanagement issues.
20. Evaluate any problems that may be associated with gross negligence or willful misconduct. Obtain guidance from legal counsel on these issues.
21. Identify any "specific information on matters of substance" that the owner may have withheld that, if provided, would have affected the contractor's bid price.
22. If differing site conditions are being alleged, obtain all available soils reports and descriptions of site surface and underground conditions that the owner provided in the bid package and contract documents. Also, obtain all reports of the actual site surface and underground conditions.
23. If weather-related delay claims are being alleged, obtain all reports relating to weather conditions from the project record and from the closest weather station to the project site.

19.2 Claims Analysis Related to Schedule Delay

1. Obtain all native electronic files for the baseline schedule, any rebaseline schedules, and periodic schedule updates.
2. Determine if the schedules contain logic errors or missing logic, or are governed by an unusually high number of constraints.
3. Determine if the contractor's as-planned schedule (sequences, durations, scope) was reasonable. Confirm that the as-planned schedule included the entirety of the originally contracted scope of work.
4. Identify specific requests for time extension sent to the owner and actual time extensions granted by the owner due to delays caused by approved change orders, and time extensions that are associated with pending change orders or other problems, such as *force majeure* events, suspensions, etc.
5. Identify any attempts to accelerate the work as a result of the problems encountered. Determine if the contractor was directed to accelerate or if constructive acceleration has occurred.
6. Determine if the contractor planned to complete the project work early. Determine how this early completion date was determined and how it was communicated to the owner.
7. Determine if the as-built start and finish dates for the schedule activities are accurate.
8. Identify the relevant milestones and/or time periods that must be analyzed. In some projects, for example, liquidated damages are assessed based on the delays to interim project milestones and not just project completion. Additionally, a contractor's time-related costs may be highly variable during the lifespan of a project, *e.g.*, high during the construction phase but much lower during the engineering phase.
9. Determine the appropriate retrospective schedule delay analysis method(s) that need to be performed based on the delay-related issues being alleged. Refer to AACE International's Recommended Practice 29R-03, Forensic Schedule Analysis, for selection of the most appropriate method(s).
10. Determine if the schedule analysis should employ an assessment of delays during specific time periods (*i.e.*, windows), such as between monthly schedule updates.
11. If appropriate and necessary, correct the schedules (logic, dates, durations, etc.) before performing a retrospective delay analysis, depending on the schedule delay analysis method(s) being performed.
12. For certain schedule analysis methodologies, prepare activity duration and relationship variance tables to quantify start and duration delays to schedule activities. These variance tables may need to be calculated for each of the various time periods being analyzed.

13. Apportion delays to the activities identified in the duration and relationship variance tables. Determine responsibility for delays using the claim issue analyses (Section 2 above) already performed and seeking guidance from legal counsel and/or technical experts if appropriate.
14. Further analyze any large delays identified in the variance tables that the claim issue analyses do not explain.
15. Identify the planned and actual start dates, milestone and final completion dates, and activity durations for activities associated with the problems, as well as activities on the critical and near-critical paths.
16. Determine if the as-planned and as-built critical paths are different for each period of the analysis.
17. Identify specific critical and near-critical activities that were delayed and/or impacted by the problems encountered.
18. Determine if the work affected by the problems was (1) on the as-planned critical path, or (2) on the statused critical path at the time the problem occurred.
19. Determine if there were other delays that occurred before or after the problems that may be related to or were caused by the problems, *i.e.*, dependent concurrent delays.
20. Identify any independent concurrent delays specific to the time frame of each problem.
21. Determine if the work was performed out of sequence as compared to the as-planned schedule.
22. Determine if there are any compensable delays associated with differing site conditions, suspensions, work being delayed into bad weather periods, or other claim issues for which time extensions have not been requested and granted.
23. For certain schedule analysis methodologies, prepare schedule delay fragnets based on the identification of the various delays (from claim issues analysis described in Section 2 above), and identify the existing schedule activities that were impacted. A fragnet is a group of schedule activities inserted into a schedule that model delaying events.
24. Use an appropriate retrospective schedule analysis, such as a time impact or update impact analysis, to determine if the contractor is entitled to a time extension.
25. Use an appropriate retrospective schedule analysis that considers concurrent delays, such as an As-Built But-For Analysis, to determine if the contractor is entitled to compensable delay time and costs.
26. Use an appropriate retrospective schedule analysis to determine if the contractor accelerated the work and use the results of this analysis to

allocate acceleration costs that have not already been paid through change orders.

27. Use an appropriate retrospective schedule analysis to determine if the owner is entitled to liquidated or actual delay damages.
28. Prepare narratives with supporting documentation that describe the information utilized, the problems encountered, the analyses performed, and the results of the schedule calculations and results.

19.3 Claims Analysis Related to Quantities, Man-Hours, and Costs

1. Obtain and review the contractor's bid and bid estimate calculations and assumptions, including productivity, material quantities, labor rates, contingency, field and home office overhead amounts, and profit.
2. Determine if the bid was reasonable, or if there could have been an underbid or bid error.
3. Obtain and review the contractor's budget and revised budget information, including quantity, man-hour, and cost data and underlying assumptions, noting any differences from the bid estimate.
4. Obtain and review the contractor's actual quantity, man-hour and costs data.
5. Determine the extent to which the contractor coded its additional man-hours and costs to discrete change-related cost codes, *i.e.*, did the contractor segregate its additional work and its base scope work?
6. Obtain and evaluate the contractor's work breakdown structure (WBS) coding and/or cost account structure. Evaluate the level of detail to determine if (disruption) claims can be assessed by discipline, work area, timeframe, etc. Confirm that the same coding structure was used for budgeted and actual man-hours and costs.
7. Obtain and evaluate the approved, pending, and disputed change orders on the project. To the extent possible, determine the material quantities, man-hours, and costs included in the changes (which may be difficult in respect to negotiated, lump-sum change orders). Determine if the approved change orders included field and home office overhead costs.
8. Determine if accord and satisfaction was reached in the approved change orders. Are claims being sought for problems that have already been resolved? Did the contractor reserve its rights to assert additional claims?
9. Prepare a quantity variance analysis by material type or detailed WBS/cost account to determine which types of materials had significant variances when planned and actual quantities are compared, including consideration

of any approved or pending change orders. Determine the extent to which quantity overruns affected labor man-hours and costs.

10. Prepare a man-hour variance analysis, comparing budgeted and actual man-hours at the most detailed level of the information available, including consideration of any approved or pending change orders.
11. Prepare a cost variance analysis, comparing budgeted and actual costs at the most detailed level of the information available, including consideration of any approved or pending change orders.
12. In preparing a claim, utilize the variance analyses to: (1) identify and claim discretely coded additional work; (2) identify accounts with overruns that are claimable; (3) identify accounts with overruns that are not claimable; (4) identify escalation costs to labor, materials, or equipment, based upon comparisons between budgeted and actual rates; (5) quantify the claimed amounts for certain cost accounts when a Total Cost or Modified Total Cost Approach is used; and (6) demonstrate that the claimed costs are reasonable, by showing the claimed amounts as compared to the variances or losses.
13. In defending against a claim, utilize variance analyses to: (1) identify potential underbid or bid error; (2) demonstrate that claims are being improperly asserted for accounts in which no man-hour or cost overrun was experienced; (3) demonstrate that the entirety of the man-hour or cost overrun for a given account is being claimed improperly because of demonstrable contractor-responsible problems; and (4) identify contractor-responsible issues (*e.g.*, material quantity growth from its budget) that may be used to demonstrate incorrect claim amounts.
14. Related to claims for additional work: (1) confirm contemporaneous estimates included in change order documentation; (2) evaluate and confirm additional costs using invoices and/or labor man-hours and costs from discrete change-related accounts; or (3) alternatively, when better data is not available, estimate the extent of the additional work.
15. Prepare a disruption or loss of productivity claim, which is often the largest component of a contractor's claim. To the extent possible, compare the planned working conditions with the less-productive actual working conditions for discrete and measurable scopes of work, provide the causes of the changed conditions, and quantify the resulting loss of productivity.
16. Obtain if available contemporaneous data with respect to labor man-hours incurred and quantities installed by time (week or month). In preparing a disruption claim, utilize such time-series data to demonstrate productivity losses in discrete periods or show work stoppages or out-of-sequence work. Alternatively, in defending against a claim, determine if the timing and magnitude of the claimed man-hours align with the alleged impacting events.

17. If appropriate man-hour and installed quantity data is available over the duration of the project, and if there are relatively unimpacted periods of work, perform a measured mile analysis to determine the loss of productivity man-hours that may be associated with the alleged impacts.
18. Alternatively, utilize various industry studies to estimate the loss of productivity man-hours associated with impacting events, such as:
    - Overtime
    - Stacked trades/congestion
    - Weather impacts
    - Out-of-sequence work
    - Cumulative impact due to multiple change orders
    - Other productivity impacts
19. Determine if management staff or overhead costs increased as a result of disruption, so that such discrete costs could be claimed as part of a disruption claims (rather than as part of a delay claim).
20. Determine if the contractor expended additional costs for acceleration and if the owner has paid for any acceleration costs.
21. Determine if the change orders resulting from the alleged problems included impact or loss of productivity costs.
22. Related to prolongation or delay claims, quantify the contractor's average time-related field overhead costs by day, week, or month. Use the results of the schedule analysis for compensable delay to quantify the contractor's delay damages. As noted above, the contractor's overhead costs may change substantially during the project, and more than one period may have to be considered.
23. Determine if there are any specific home office overhead man-hours and costs that can be tied directly to the delays. Consider the use of the Eichleay or other overhead cost allocation formulae to calculate the compensable home office overhead delay costs.
24. Determine what construction equipment was used for work relating to the problems. Obtain and analyze the type of equipment, duration of the equipment's use, and costs associated with construction equipment ownership/rental costs during the period of delay. Using the delay analysis results, apportion construction equipment delay costs that were caused by the owner and caused by the contractor.
25. Using the man-hour and cost variance analysis, identify any labor, materials, or installed equipment escalation costs compared to the bid estimate for labor, materials, or installed equipment due to delays caused by problems.
26. Update the man-hour and cost variance analyses to include approved and pending change orders, contractor-caused problems, delays, and other claim issues to determine if the amount being claimed at the cost account

level is reasonable. Ensure that the man-hours and costs associated with any claims for which the contractor may be entitled fit into the man-hour and cost variance matrices such that the claimed man-hours and costs do not exceed the difference between the current contract value and the actual man-hours and costs when contractor-caused problems are accounted for in the variances.

27. Calculate the reasonable claim value associated with unresolved suspension or termination claims.
28. Calculate the reasonable value of any lost profits claims.
29. Calculate the reasonable value of any misrepresentation claims.
30. Calculate the reasonable value of any claims associated with a failure to perform to a standard of care.
31. Calculate the reasonable value of any impossibility of performance claims.
32. Calculate the reasonable value of any unjust enrichment claims.
33. Calculate the reasonable value of any superior knowledge claims.
34. Calculate the reasonable value of any breaches of implied warranties claims.
35. Calculate the reasonable value of the contractor's claim and the owner's counterclaim against the contractor.
36. Prepare cause-effect matrices that connect the root causes of the problems to intermediate causes and then to the ultimate effects and costs.

Prepare narratives with supporting documentation that describe the information utilized, the problems encountered, the analyses performed, and the results of the damages calculations.

Index

Index

Index

Case Citations

www.ingramcontent.com/pod-product-compliance
Lightning Source LLC
Chambersburg PA
CBHW041257150926
51501CB00012B/215
9781638680161